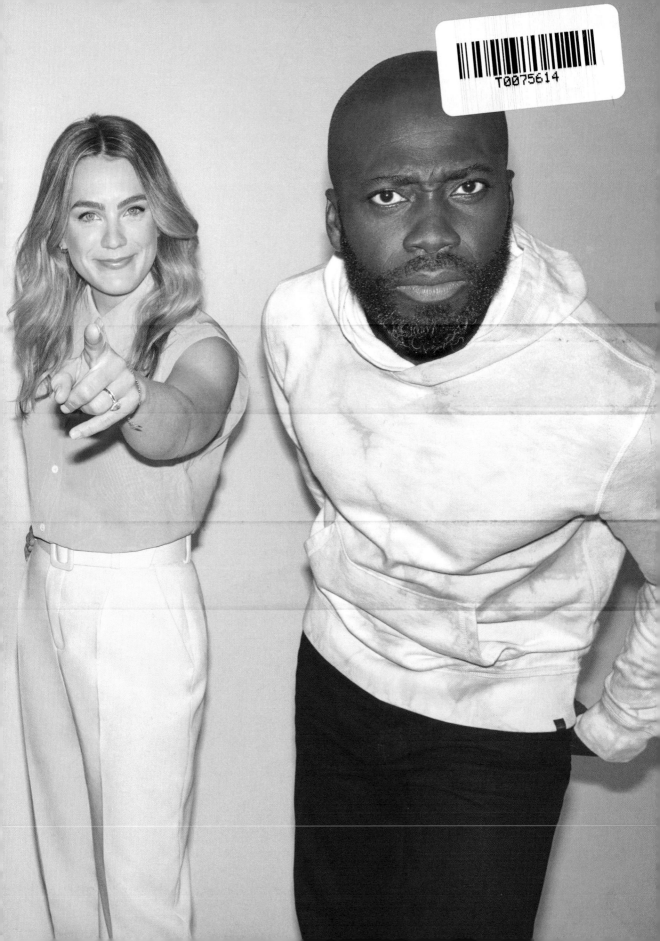

The Gadget Show

BIG BOOK OF COOL STUFF

Edited by Caramel Quin

black dog press

Welcome

On *The Gadget Show*, we love all things tech. Since we hit the screens in 2004, we have bounced up and down excitedly in our front-row seats as technology has transformed the world. The stuff of science fiction is increasingly a reality (page 120). The robots are rising (page 158). And while Jon tells us not to hold our breath for flying cars like Doc Brown's DeLorean, he assures us that lightweight passenger drones will be a reality in the near future (page 10). We believe we can fly.

We've all had our fair share of adventures presenting the show (page 42). It's an absolute joy getting to play with tomorrow's techie toys, and in this book we've tried to share that passion with you. Not only showing you the wild and wonderful world of gadgets, but also the innovative bits of tech that transform everyday life, making it easier and more enjoyable.

Lockdown proved how important technology is. It can connect us with loved ones, support home working (page 134) and home schooling, or help us cook up a storm like a professional chef (page 148). Gadgets can even transform our homes (page 126) to make them smarter and do our bidding effortlessly. We can increasingly control them with our voices. Thanks to artificial intelligence, our digital assistants understand and can talk back or translate languages instantly, connecting us with others (page 172).

We've cherry-picked the best new technology for you and, as Craig says, kicked the stuff that's "tech for tech's sake" (page 36) to the kerb. So welcome to *The Gadget Show Big Book of Cool Stuff*. We had great fun putting it together. We hope you have just as much fun reading it.

The Gadget Show team

CONTENTS

MEET THE TEAM

26

FUN AND GAMES

56

MIND AND BODY

86

CO

102

104

110

NTENTS

IN THE HOME

148

CONTENTS

A vision of the
future, inspired by
Blade Runner 2049

DUDE, WHERE'S MY FLYING CAR?

THE GADGET SHOW'S RESIDENT PETROLHEAD
JON BENTLEY LOOKS TO A FUTURE BEYOND FOSSIL
FUELS, BEYOND DRIVERS... AND BEYOND ROADS

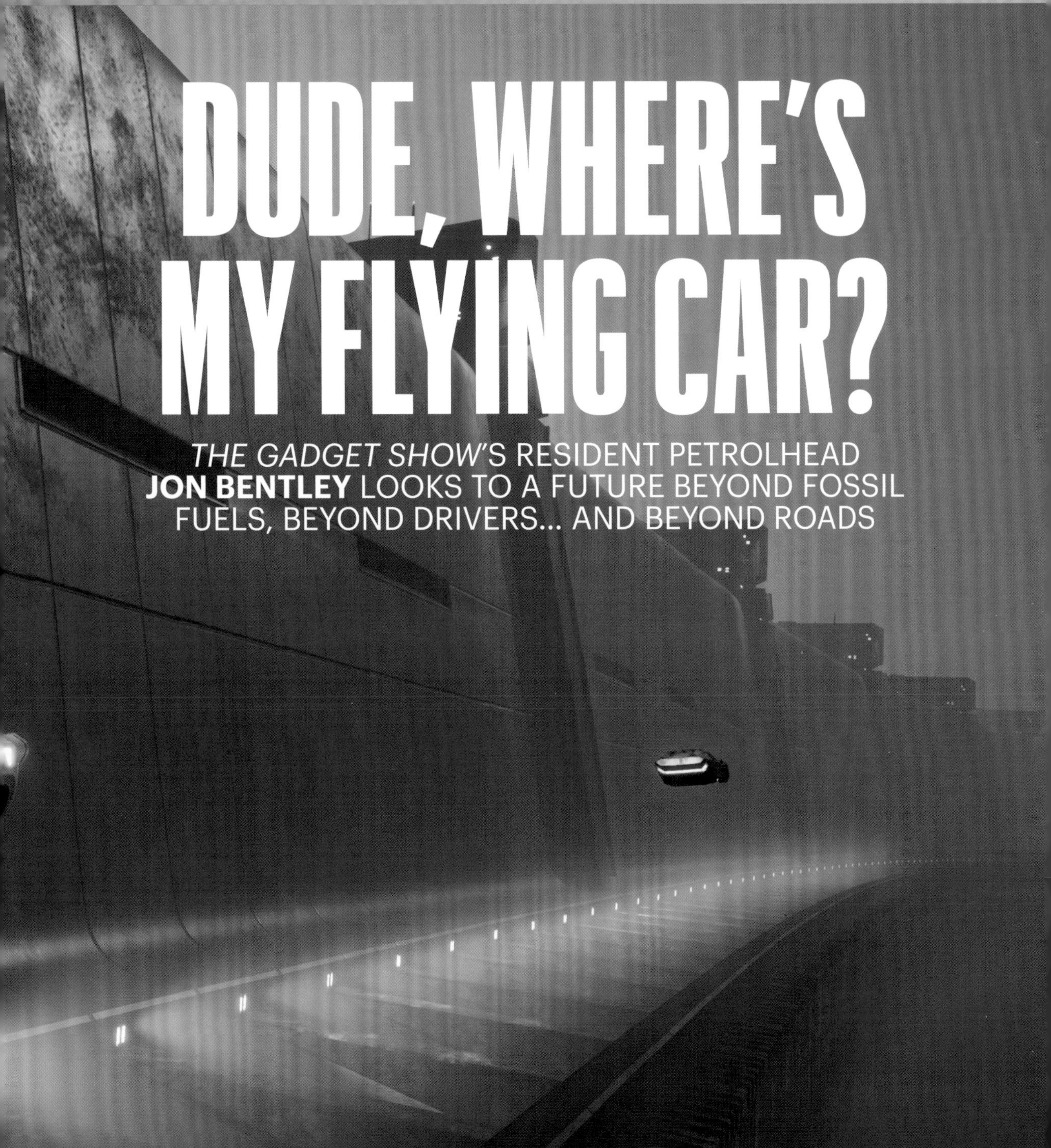

Being known as something of a car enthusiast, conversation often turns to the future of motoring, and the first thing everyone asks is, "When will we have flying cars?" The question is baked into the public imagination, and prototypes are nearly as old as the car itself, the first appearing as long ago as 1917 when Glenn Curtiss unveiled his Autoplane. Fiction is full of them, from *Chitty Chitty Bang Bang* to Doc's DeLorean in *Back to the Future Part II* to the Spinner police cars in *Blade Runner*, but they are yet to become a practical reality.

Unfortunately, that fact won't change over the next few decades, but many other aspects of cars will. The first, already well underway, is the transition to new sources of power, spurred on by government pledges to ban the sale of new petrol and diesel cars (as early as 2025 in the case of Norway and 2030 in the UK). The weapon of choice in the battle to decarbonise personal transport? The lithium-ion cell, a technology you already

know and love for its ability to power our phones, laptops and other favourite gadgets.

As the cost decreases and performance gradually improves, these rechargeable batteries are the current go-to for electric vehicles. When they are mated with an efficient electric motor, the result is quick, quiet and easy-to-drive cars with no tailpipe emissions. At the highest performance levels, cars such as the Rimac Nevera and the 2022 Tesla Roadster – the latter expected to be capable of 0–60mph in 1.9 seconds, with a top speed of over 250mph – are proving that electric cars are more than a match for internal combustion-engine supercars in terms of maximum speed and acceleration. Meanwhile, Volkswagen's ID series, the Tesla Model 3, Hyundai's Ioniq 5 and the Porsche Taycan are banishing any lingering association that battery power may have with milk floats and box-like city cars. The race is on to build more charging stations.

However, lithium-ion batteries do come with their own set of problems and aren't quite the super-saviours we would like them to be. Not only are they expensive, heavy, difficult to recycle and slow to charge, but the environmental and humanitarian cost can be extensive – a sad irony considering the push towards electric vehicles is

Tesla's all-electric Roadster, with a top speed of over 250mph and a range of 620 miles

Right: Doc Brown's DeLorean time machine in *Back to the Future Part II*

Far right: Star of the film *Chitty Chitty Bang Bang* – a flying vehicle inspired by a 1920s racing car

meant to offset the impact of fossil fuels. From the CO_2 emissions embedded in their manufacture to toxic leaks associated with the mining process and the fact that the raw materials needed to make them are mined and processed by people working in unsafe conditions in unstable economies, the pitfalls must be weighed against the positives.

As a result, Tesla has already announced that it will stop using cobalt to make the cathodes in its electric vehicle batteries and that its move towards better battery technologies will mean less of these problematic materials are needed. Other carmakers such as Toyota and Nissan are developing cars powered by solid-state batteries (the material between the battery's two electrodes is a solid, not a liquid) for longer ranges and faster charging. Expect them to appear between 2025 and 2030. One revolutionary new solid-state technology making headlines is the glass battery, devised by materials scientist (and inventor of the lithium-ion battery) John Goodenough, which uses a glass electrolyte and lithium or sodium electrodes. Goodenough claims they have an energy density many times that of current batteries, a longer life and much shorter charging times. But don't hold your breath – they are currently far from being proven, tested or scaled up to practical levels.

Supercapacitors – which unfortunately do not share the same time-travelling capabilities as Doc's *Back to the Future* flux capacitor – are an exciting potential

new power source, too. They bridge the gap between electrolytic capacitors and rechargeable batteries, and they can absorb and release power extremely quickly. The trouble is, their energy density is pretty rubbish. Scientists are exploring the use of polymers similar to those used in disposable contact lenses in order to solve this problem. The much-hyped supermaterial graphene (see page 184), with its huge surface area, could form the structure of future game-changing supercapacitors. In fact, supercapacitor body panels, as seen on Lamborghini's Terzo Millennio concept

YOUR CAR COULD BECOME CARBON NEUTRAL BY FILLING UP THE TANK WITH SYNTHETIC FUEL, AN ECO-FRIENDLY REPLICA OF PETROL OR DIESEL

supercar – a collaboration with MIT – may yet make it into production.

Meanwhile, your current car could easily become carbon neutral by filling up the tank at the petrol station with synthetic fuel, which is effectively an eco-friendly replica of petrol or diesel. Made by combining CO_2 from the atmosphere with hydrogen from water, synthetic fuels have a closed loop because the CO_2 used to produce them would equal the amount pumped into the air by cars burning it, making it CO_2-neutral to run. However, as a result of high production costs, synthetic fuel – due to be made in giant processing plants erected in parts of the

Lamborghini's Terzo Millennio concept supercar, with supercapacitor body panels

world with surplus solar energy – will most likely be earmarked for use in aircraft first.

Slightly closer to becoming a ubiquitous energy source is hydrogen itself. At present, most hydrogen is produced from natural gas, meaning it isn't carbon neutral. (Those 'zero emissions' hydrogen-fuelled London buses actually have to read, "zero emissions at the tailpipe" in the small print, with tailpipe emissions consisting of water alone – a huge plus for air quality nevertheless.) But in the future our hydrogen could be far greener. Plentiful renewable electricity could be used to create hydrogen by splitting water, making it an excellent carbon-neutral fuel for cars. Used to generate electricity in fuel cells, as in the Toyota Mirai, hydrogen gives a 400-mile-plus range and you can fill up in roughly the same amount of time it takes to fill a petrol tank. Although expensive for now, Toyota reckons hydrogen-powered cars will cost no more than a current hybrid by around 2030.

ROBOCAR TO THE RESCUE

When it comes to the depiction of futuristic worlds in popular culture, self-driving cars feature nearly as much as their flying counterparts. While there is still a long road ahead until we are reading books, playing video games or taking naps in the back of our car as it cruises safely to our destination, self-driving technology promises to change our whole notion of the car. With no need for a steering wheel and controls, in theory, any living space could be on autonomous wheels – from pubs and hotels to doctors' surgeries and theatres. I'm personally very excited by the idea that my future house could have a mobility room that detaches and travels about.

In reality, true self-driving is currently undergoing a bit of a reality check. The sensing technology is good: radar, sonar, cameras and lidar, which uses a pulsed laser to measure ranges, all contribute to cars being very aware of their environment. The challenge is developing the artificial intelligence (AI) necessary to interpret and respond to the colossal amounts of information they generate (by some estimates up to 5GB of data a second). It's the biggest test of AI that humanity has ever come up against. Elon Musk promised his Tesla cars would be fully self-driving by 2018. They weren't. This is just one of many missed deadlines in the market.

For now, the tech is being tested in robot taxis. Waymo – a subsidiary of Google's parent company Alphabet – has rolled out a fleet of state-of-the-art rides, which are some of the world's only fully driverless vehicles operating on public roads today. But even these 'rider only' cars have a remote driver at HQ ready to push a button at any time and put the car back on course. They struggle on the grid-like road networks of Phoenix, Arizona (and would likely have difficulty on *Tron*'s Grid). Fellow road users quickly lose patience with their stuttering, cautious progress, and some especially bold drivers have been playing chicken by driving head-on towards autonomous cars; a few unfortunate occupants have even experienced guns being fired at their robot taxi.

SEMI-AUTOMATION

A slightly more realistic goal for the near future is semi-automation – the idea that cars can be left to their own devices in slow-moving traffic so you can concentrate on reading or answering emails, but when speeds increase the car will signal that you need to snap back into control. My prediction is that this will likely prove too dangerous, as I think many drivers will be unable to switch modes quickly or reliably enough. Self-driving cars will also communicate with one another to reduce traffic casualties, but these more connected cars will bring bigger cyber security challenges. The thought of hackers taking control of thousands of cars at once makes me shiver. More positively, these developments in accident-avoidance technology, together with electric platform chassis, may remove the need for type approval and crash-testing.

The combined effect of these hurdles means that self-driving will likely roll out in locations where the environment can be strictly controlled. At the slow end, this will be university campuses or retirement complexes where pod-like vehicles can crawl around in safety, while small areas of cities or inter-urban corridors will allow for slightly higher speeds. At the fast, more exciting end, it will be racetracks where you can be taught to drive at high speeds by a

WE MIGHT CHERISH OUR HISTORIC CAR, USE AN ELECTRIC VEHICLE FOR URBAN TRIPS, THEN TAKE SOMETHING MORE AUTONOMOUS WITH A HYDROGEN POWER UNIT FOR LONGER JOURNEYS

virtual Lewis Hamilton – I really can't wait for my first lesson.

In a joyous throwback to the days of coachbuilding, in the future you will be able to commission a bespoke car. Just discuss your preferred option with a designer, draw up a sketch on an iPad, then 3D print the parts and have the haute couture car delivered to your home in a matter of weeks.

To keep classic cars in production, 3D printing will also help with the manufacture of discontinued parts. Classic cars, like steam trains, have such a tiny contribution to pollution that they will continue to be exempt from tightening environmental

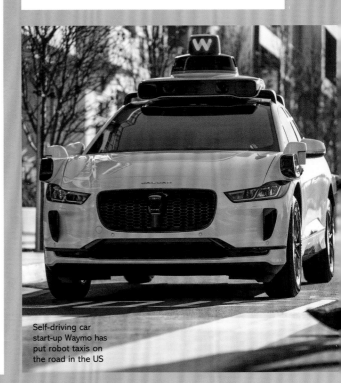

Self-driving car start-up Waymo has put robot taxis on the road in the US

legislation. There will be increasing pressure to list certain classic cars as works of art, like historic buildings, and this has already started in America with cars such as one of the two Ford Mustangs driven by Steve McQueen in the film *Bullitt* listed on the National Historic Vehicle Register.

Different countries will demand different types of car. America's enthusiasm for pick-ups will be bolstered by electric offerings such as the Rivian R1T, Tesla Cybertruck and Ford F-150 Lightning. As with smartphones, Chinese brands will become more familiar and develop their own distinctive designs, such as Geely, BYD, Xpeng and NIO. Some may even be smartphone brands: Xiaomi is exploring the world of cars, for example. Africa is set to be the world's fastest-growing car market in the 2030s with its own design demands, and other new brands will stimulate more creative design as they develop their own identities.

The urban versus rural split in our attitudes to personal mobility may deepen, too. Urban transport will be increasingly public or two-wheeled, with better battery power helping the adoption of bicycles and scooters. There will be less car ownership, and more car sharing and autonomous vehicles in dedicated areas. In rural and most suburban areas, however, people will continue to own and enjoy driving their cars. Attempts to cut car ownership through nanny-ish planning and punitive taxation could even make it seem more prestigious and desirable.

Wherever you live, future design is an inescapable fact, and we will embrace a wider range of cars. We might cherish our historic car and be able to exercise and maintain it on a heritage campus, use an electric vehicle for urban trips, then take something more autonomous with a hydrogen power unit for longer journeys. We will pay subscriptions to motor sport driving experience parks where AI and virtual reality (VR) offer unparalleled instruction on how to drive better. Cars will be as beautiful and desirable as before, representing freedom and some of the best tactile pleasures that life can offer.

What about flying cars, though? Well, just as today's 'hoverboards' don't actually hover, we will soon have the flying cars we were promised… they just won't really be cars.

Classic cars as works of art: one of the two Ford Mustang GTs driven by Steve McQueen in the film *Bullitt*

Joby Aviation's all-electric aircraft prototype – an air taxi of the future

I BELIEVE I CAN FLY

WILL PASSENGER DRONES TAKE OVER FROM THE FANTASY OF FLYING CARS?

In a sci-fi and superhero world, there would be no need for cars – we would all use jetpacks. Sadly, they aren't set for mass adoption. The need for operator skill and training, plus vast cost and high noise levels, limit them to the extremely wealthy in search of a few minutes of exhilaration. JetPack Aviation's JB11, with six kerosene-powered turbojet engines, is believed to cost $340,000.

Meanwhile, *Back to the Future Part II*-style hoverboards will remain technically challenging, to say the least. Ironically, the closest I have seen is a fantasy project by a car manufacturer. The 11.5kg Lexus Slide runs on a magnetic track and the board uses liquid nitrogen to cool superconducting blocks to -197°C, at which temperature they produce a maglev-like opposing magnetic field that makes the board hover.

That long-held dream of personal transport, the flying car, will stay frighteningly expensive and surprisingly inconvenient. Take-off normally requires a runway, and you would need a pilot's licence on top of your normal driving licence. It would also tend to be a bit rubbish at being both a car and a plane. A car must be strong for crash resistance, with large tyres to aid roadholding, whereas a plane needs to be light with tiny wheels to get off the ground, and it has traditionally required bulky wings, which rather get in the way when negotiating terrestrial traffic. It's almost impossible to design a single vehicle that is good at both. But that doesn't stop people trying. Spectacular recent concepts include the Terrafugia TF-X, which promises a 500-mile range and autonomous capabilities, and the rather stylish AirCar prototype from Slovakia, which can transform from car to plane in 2 minutes and 15 seconds.

JetPack Aviation's JB11 comes with six kerosene-powered turbojet engines and a $340,000 price tag

The future of personal flying transport looks more likely to be small electric vertical take-off and landing (eVTOL) craft and passenger drones. If anything, just using the term 'flying car' labels you as a dinosaur. Prototypes being developed include Joby Aviation's S2 and Volocopter's VoloCity. South Korean car company Hyundai is working with Uber to develop an eVTOL taxi service that it hopes to launch by 2030. But these still need some improvement in battery energy density and will be strictly the preserve of a minority. En masse they take up far too much air space. Imagine everyone leaving a train station and hopping into their drones. The sky would be black with a swarm of buzzing activity. Also, yet to be tested is whether the sensation of being whisked around in a drone is too close to a manic rollercoaster ride to be acceptable as transport.

Gadgets on the go

Wheelie good technology for drivers, motorbike riders and cyclists

ORTLIEB ATRAK 25L
This waterproof bag really must be seen. It's a backpack when you're on the move, but undo the long, waterproof zip on the spine side and it opens like a duffle bag. It's easy to find items inside, and easy to pack.

WAHOO ELEMNT BOLT

A GPS bike computer for the cycling enthusiast who loves data as much as café rides. Its 2.2-inch colour screen displays precisely the route directions and data you need, and pairs with cycle sensors and your phone.

GROOV-E GV-WMG

Quite the most gadgety in-car phone mount, this offers wireless charging and has a clever infrared auto-grip system that detects and adjusts to the size of your device. Mount it on the windscreen or an air vent.

GARMIN RALLY RS200

These high-end smart pedals quantify every element of your pedalling. They work with SPD-SL cleats and can be used with any bike. They can also tell you how much of a workout you're getting on an e-assist bike.

GARMIN DASH CAM 57

This small cam sticks to the car windscreen and can record in HD constantly with GPS stamp in case of an accident. It also features collision warnings and voice control and can monitor activity around your car when parked.

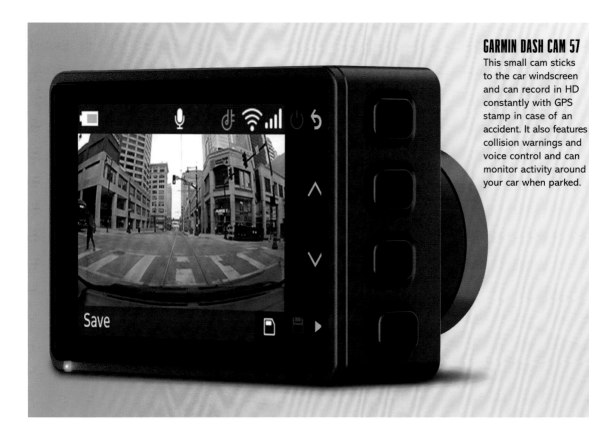

BLUEAIR CABIN P2I

If you're stuck in a traffic jam, you're stuck in polluted air, even if your own vehicle is green. Enter the Cabin P2i, which filters the air in less than six minutes, removing particulates, pollen, pet dander and more.

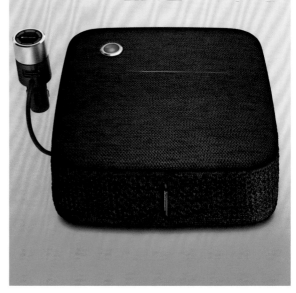

CROSSHELMET X1

The only motorbike helmet with twin heads-up displays (HUDs). See essential riding information including directions, weather and time, as well as a rear view from a wide-angle camera built into the back.

HÖVDING 3

An innovative airbag for cyclists that's built into a collar. If you're in a collision it inflates in 0.1 seconds, giving eight times more protection than a helmet. It can trigger your phone to call your emergency contacts, too.

SMARTHALO 2

This elegant cycle computer clips to the middle of your handlebars and displays turn-by-turn directions. It also has a built-in front light and an alarm. If you forget where you left your bike, find it with the app.

ALPINE HALO9 ILX-F903D

Unusually for big in-car entertainment upgrades, the iLX-F903D fits in a single DIN (the size of a small car hi-fi). Its large touchscreen stands proud of the dashboard for Apple and Android screen mirroring, USB video and more.

Starman waiting in the sky

Is it a bird? Is it a plane? No, it's a sports car orbiting the sun!

Love or loathe him, Elon Musk knows how to do a publicity stunt that's out of this world. So, when his SpaceX heavy-lift launch vehicle Falcon Heavy needed a trial payload for its first launch, he sent a Tesla Roadster electric sports car into space, complete with a mannequin in a spacesuit (pictured below). Musk is a huge fan of Douglas Adams's *The Hitchhiker's Guide to the Galaxy*, so there was a copy in the car's glove box along with a towel, and the words 'Don't Panic!' on the dashboard. The car is now in orbit around the sun. The current space race between billionaires is set to transform travel for the rest of us. A sub-orbital flight could mean London to Sydney in less than an hour.

What a carry on

Gadgets to help you travel in style for business or pleasure, whether by plane, train or automobile

1 DYSON CORRALE
This go-anywhere cordless hair straightener has copper flexing plates that fit around the section of hair to be styled, requiring less heat. This reduces hair damage, frizz and flyaways. One charge provides 30 minutes of styling.

2 HYLETON TRAVEL ADAPTOR
Charge all your gadgets with this adaptor for use worldwide. It takes one mains plug but boasts four USB ports plus a higher power USB-C.

3 BELLROY TECH KIT
Gadget-loving travellers need a way to stow tech without creating a tangled mess. This is big enough for your charger, mouse, headphones, USB sticks and all the cables. There's also a flat pocket for a power bank.

4 RAVIAD MULTI CHARGER CABLE
A travel lifesaver, this USB cable is triple-ended with Micro USB, Type-C and iPhone Lightning connectors. You can use all three at once. The life you save – or the unexpected phone you charge – might not be yours.

5 HAMPTON BENJILOCK
Add biometric security to your luggage with this fingerprint-sensor padlock. It stores up to five fingerprints. There's no key but it is compatible with TSA (Transportation Security Administration) keys for US travel.

6 INSTA360 GO 2
This tiny action cam can be worn, carried or mounted, to vlog every moment. Attach it to its magnetic pendant to capture POV (point of view) videos effortlessly. The charge case doubles as a remote control and tripod.

7 SALTER LUGGAGE SCALES
Lift your suitcase using the Salter's handle for a quick weigh-in to avoid excess luggage charges and airport stress. There's just one button and a simple LCD display.

8 VODAFONE CURVE
An affordable smart tag with built-in SIM card, GPS, Wi-Fi and Bluetooth. For a small monthly subscription, you can find your luggage (or handbag or pet) anywhere, without relying on a phone to be nearby.

9 POCKETALK S
Speak for up to 30 seconds in any of Pocketalk's 82 languages and it automatically translates. You can type in text or point its camera at signs, too. It comes with two years of free data in more than 130 countries.

10 BOSE 700
Bose is well known for its superb noise-cancelling headphones for travel. They are big and comfy, so you won't tire of them on a long journey and their 20-hour battery life will go the distance, too.

11 SAMSONITE C-LITE
This 55cm carry-on case is made from a strong yet light, woven polypropylene, called Curv, so it can take the knocks. An integrated USB port next to the lock charges gadgets and the TSA lock is good for US travel.

12 EKSTER ALUMINUM CARDHOLDER
This aluminium wallet fans out cards at the touch of a button and protects from RFID wireless theft. Add the solar-powered Tracker Card (pictured) to find your wallet via an app or voice command.

13 LIGHT MY FIRE SPORK'N STRAW
This Swedish-made, titanium spork (spoon and fork combined) and straw are strong, lightweight and won't corrode. They come in a recycled merino wool case, perfect for your travels and for dodging single-use cutlery everywhere.

14 KINDLE OASIS
This ereader has a crisp 7-inch screen with an adjustable light; choose to go whiter or warmer. It weighs just 188g but can hold thousands of books electronically. It's waterproof, too.

Chapter One

14

13

12

11

6

10

7

8

9

FAST v

This photo: Rent a cabin on a cargo ship as a slow and sustainable way to cross the seas

Opposite: Candela C-7, an electric hydrofoil speedboat with a battery-driven range of around 57 miles

SLOW

THE FUTURE OF MASS TRANSIT IS GOING TO BE SUPER FAST... OR SUPER SLOW. GET THERE QUICKLY OR ENJOY THE JOURNEY IN LUXURY

FAST: HYPERLOOP

TAKE THE HIGH-SPEED TRAIN THAT TRAVELS INSIDE A VACUUM TUBE

Tesla CEO Elon Musk proposed Hyperloop in 2012. A Hyperloop train travels inside a tube that's a near vacuum, to reduce drag so the train can travel faster. Air bearings or magnetic levitation (maglev) lift the train from its rails to further cut friction.

In his 2013 white paper 'Hyperloop Alpha', Musk presented a system that could move passengers and cargo at 760mph, going from Los Angeles to San Francisco in 35 minutes. Solar panels on top of the tubes would be more than enough to power the trains.

Musk's company SpaceX built a working-scale model in California. So far, its world record is 288mph. Meanwhile, Richard Branson's company Virgin Hyperloop was the first to conduct a human trial. Its 500-metre Las Vegas development loop took its first Hyperloop passengers at a speed of 107mph. These vacuum tube trains, or vactrains, can potentially go under water, raising the prospect of a high-speed transatlantic tunnel.

SLOW: SLEEPER TRAINS

THE NOSTALGIC LUXURY OF OVERNIGHT RAIL TRAVEL

The future of rail travel could equally mean slower journeys in comfort, with sleeper trains that evolve from the nostalgic opulence of trains such as the Orient Express and Trans-Siberian Express.

If you're travelling for work you can arrive rested, having eaten a five-star meal in the dining carriage, slept and showered on the train. You could even indulge in the old-fashioned luxury of reading a newspaper over breakfast. Private cabins with coffee on tap and Wi-Fi would allow the peace and quiet you need to work on the move. Holidays could be just as relaxing: enjoy the views from the window and unwind before you reach your destination.

French start-up Midnight Trains plans a network of sleeper services out of Paris from 2024. One of its backers is billionaire Xavier Niel, telecoms entrepreneur and co-owner of *Le Monde* newspaper. The company proudly sells its 'hotel on rails' as environmentally friendly, being responsible for considerably lower carbon emissions than the equivalent journeys made by plane.

FAST: HYDROFOILS

A SPEEDBOAT THAT APPEARS TO FLY OVER WATER THANKS TO A HYDROFOIL

Like Hyperloop, the design challenge with high-speed sea travel is a drag race. The less friction the better. The Candela C-7 is an electric hydrofoil speedboat that looks like it's flying above the water. It's designed for long-range trips. Submerged hydrofoils work like the aerofoil shape of aircraft wings: as the boat gains speed, water flows around the hydrofoils creating a force that lifts the boat out of the water, dramatically reducing drag.

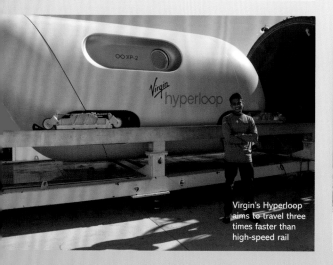

Virgin's Hyperloop aims to travel three times faster than high-speed rail

Dine in style and watch the South African scenery pass by on a Rovos Rail train

The C-7 can travel for two and a half hours at a speed of 22 knots (25mph) without creating a wake, noise or emissions. Its top speed is 30 knots (35mph). From a 12-inch touchscreen beside the steering wheel, you can access all controls and retract the hydrofoils.

SLOW: CARGO SHIPS

A GREEN ALTERNATIVE TO CRUISES

Friends of the Earth has declared taking a cruise "more harmful to the environment and human health than many other forms of travel" because of fossil fuel use even when docked, water pollution and sewage, but it sets out how clean cruising could be possible. Meanwhile, a good slow option is to rent a cabin on a cargo ship that's crossing the sea anyway.

They have cabins and better facilities than you might imagine. Don't discount actual sailing boats either. By definition, they use renewable power. When environmental activist Greta Thunberg crossed the Atlantic for a climate conference in 2019, she considered cargo ships but ultimately travelled there in a racing yacht (15 days) and returned to Europe in a catamaran (20 days).

FAST: SUB-ORBITAL FLIGHT

THE BILLIONAIRE SPACE RACE WILL LEAD TO FASTER FLIGHTS FOR THE REST OF US

The space race between Richard Branson (Virgin Galactic) and Jeff Bezos (Blue Origin) took the billionaires and their passengers into space but not quite high enough to orbit Earth, while Elon Musk (SpaceX) launched a Tesla Roadster (see page 22). Their jollies into zero-G, however, may have a pay-off for the rest of us. Sub-orbital flights would be very expensive but could mean travelling from London to Sydney in less than an hour. You'd barely have time for lunch but then, briefly in zero gravity and hurtling through the upper atmosphere, you might not be able to hold it down anyway.

Above: Virgin Galactic's rocket-powered spacecraft takes passengers on sub-orbital flights

SUB-ORBITAL FLIGHTS WOULD BE VERY EXPENSIVE BUT COULD MEAN TRAVELLING FROM LONDON TO SYDNEY IN LESS THAN AN HOUR

Below: When put into production, the Hybrid Air Vehicles Airlander 10 will operate commercial flights as a hybrid airship

Meanwhile, United Airlines proposes a return to supersonic planes, decades after Concorde was grounded. Adding Boom Supersonic Overture aircraft to its fleet will halve flight times, promising New York to London in three and a half hours. Boom Supersonic says the planes will be ready for passengers by 2029 and, in partnership with energy start-up Prometheus Fuels, use zero net carbon aviation fuel synthesised from captured atmospheric CO_2.

SLOW: BALLOONS AND BEYOND

FLY SLOW IN A LUXURY AIRSHIP OR A HYBRID PLANE

The Hindenburg disaster in 1937 ended the dream of airship travel for nearly a century. We know that it was filled with flammable hydrogen. What's less well known is that it was originally designed to be filled with helium, which is more scarce but safer.

UK company Hybrid Air Vehicles is developing a new helium-filled airship called the Airlander 10. It can carry up to 100 passengers in luxury with a range of up to 4,600 miles, staying airborne for up to five days. The company is developing hybrid electric-combustion engines to reduce emissions but, ultimately, has plans for an all-electric model.

UK start-up Faradair is developing hybrid electric planes to fly at 230mph, half as fast as current planes but they could become carbon neutral. Electric motors handle take-off while biofuels power cruising and, along with solar panels on the wings, recharge the batteries ready for landing. It's pledged 300 sustainable aircraft by 2030.

UNITED AIRLINES PROPOSES A RETURN TO SUPERSONIC PLANES ... THEY'LL BE READY FOR PASSENGERS BY 2029

FAST: AUTONOMOUS CARS

GET THERE FASTER WHEN YOU'RE DRIVEN BY ARTIFICIAL INTELLIGENCE

Self-driving, autonomous cars use artificial intelligence (AI) and immense computing power to make better driving decisions than humans. Trials right now have a human behind the wheel to take charge in emergencies but soon they could be truly

GEORGIE ON THE NEED FOR SPEED

I live in London and *The Gadget Show* is filmed in Birmingham, so a Hyperloop between the two would save me many hours. And not just for work. The pandemic has shown us that Zoom can get us so far, but human-to-human interaction is really important. It adds so much value to our lives. So, the easier it is to get about the better.

The Lyft autonomous taxi I took in Las Vegas wasn't overly cautious. When we got stuck behind a bus, it immediately checked to see if the lane was free and overtook. It was efficient.

I love that you can do whatever you want, it's your own personal pod. Imagine your family holiday: you all get in your autonomous vehicle and go to sleep. You wake up the next day and you've arrived! There's no one doing the driving, the space can be designed to fit your needs and keep you entertained.

driverless. Think of them as taxis that get you there faster and safer than cabbies. Start-ups Waymo and Lyft already offer autonomous taxis in the US that you can book via an app like an Uber.

If the cars can escape roads, they'll get you to your destination even faster. Elon Musk's imaginatively named Boring Company digs tunnels under cities for fast transport systems. Its experimental 1.7-mile loop under the Las Vegas Convention Centre allows Tesla electric cars to shuttle visitors between halls in two minutes, saving a cross-campus walk of 45 minutes or a long wait in traffic jams.

SLOW: LUXE CAMPERS

ARRIVE WELL RESTED IN THE ELECTRIC CAMPER VAN OF THE FUTURE

Electric camper vans like the VW ID Buzz and the LEVC e-Camper promise personal freedom. You take your home comforts with you: think of it as a staycation on wheels.

Nissan has even showcased a 'Winter Camper' concept, a conversion of its e-NV200 electric van with a solar panel on the roof for recharging onboard facilities. Alternatively, take it slow in the Vittra RV

(recreational vehicle), a gleaming silver machine that looks less like a bus and more like a classic Airstream caravan. In Vittra's own words it's taken the essence of an American RV and combined it with the best of European efficiency. Its 29-square-metre interior sleeps six and its fast-charging electric battery promises a range of more than 310 miles for big adventures. When it's parked at home, the van's solar roof keeps generating electricity to cut your domestic bills. What's not to like?

The electric ID Buzz, VW's prototype camper van, has AR navigation and touchpad steering

CRAIG ON LIFE IN THE SLOW LANE

I hate racing to get to places. I like the slower pace of travel. Like Will Smith says, "Two miles an hour so everybody sees you." I've spent most of the last 10 years on tour, travelling in my Mercedes Marco Polo camper van. It's got a kitchen, a shower – it's nicely kitted out. I like to set off early and get to places early, plus I can sleep all the way. My driver drives which means that I can get a bit of kip.

I'm more of a cruiser than a speed merchant, but I've never been on an actual cruise. I don't think I could get away from people. I'd love to do a cruise where I get paid to DJ though. I'd want to do it island hopping in the Caribbean.

My friend the broadcaster Ned Sherrin hated flying. He used to always get the QE2 ocean liner to travel to New York in regal splendour. It took five days when Concorde took five hours!

Hot wheels

Cutting-edge personal transport
on two or four wheels... ready to
ride into the near future

RIMAC NEVERA

An electric hypercar designed, engineered and handcrafted
in Croatia. Its top speed of around 258mph and acceleration
of 0–60mph in 1.85 seconds prove that electric cars
should no longer be seen as inferior to petrol vehicles.

SUPERSTRATA C

The Superstrata is the planet's first unibody bicycle made from continuous carbon fibre and 3D printed for a custom fit. With a 1.3kg frame, it is both light and strong.

ZERO FXS

A quiet but powerful electric motorcycle with precise handling and blistering acceleration, and a 50-mile range perfect for city living. Remarkably, you can customise the bike's performance in detail via an app.

BROMPTON ELECTRIC

This folding ebike maintains the iconic style and quality craftsmanship of the classic Brompton folding bike, but turbo charges your ride with electric assistance for speeds of up to 15.5mph.

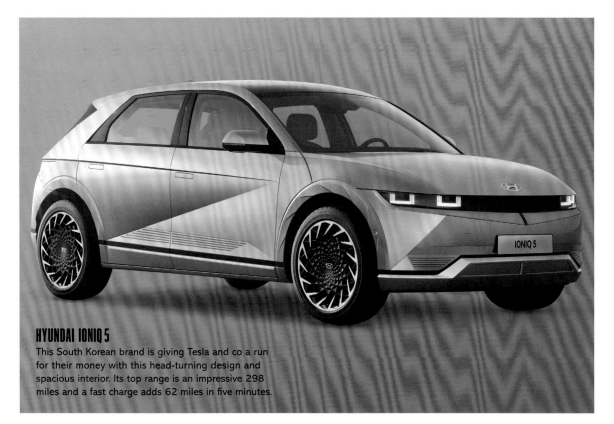

HYUNDAI IONIQ 5

This South Korean brand is giving Tesla and co a run for their money with this head-turning design and spacious interior. Its top range is an impressive 298 miles and a fast charge adds 62 miles in five minutes.

RALEIGH STRIDE 2 ECARGO

With a maximum container load of 80kg, it's good that Raleigh's cargo bike has electric assist (likewise the Pro Cargo trike, which has a maximum load of 100kg). Carry the kids and/or the shopping with ease.

CRAIG CHARLES

THE JOKER

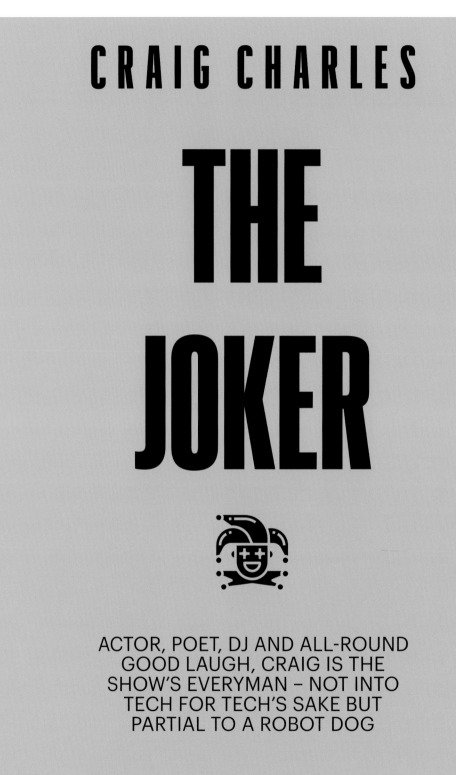

ACTOR, POET, DJ AND ALL-ROUND
GOOD LAUGH, CRAIG IS THE
SHOW'S EVERYMAN – NOT INTO
TECH FOR TECH'S SAKE BUT
PARTIAL TO A ROBOT DOG

C raig Charles's first brush with fame was jumping on stage before a Teardrop Explodes gig in 1981, aged 16, where he read out a cheeky poem which made fun of the band's lead singer, Julian Cope.

Since then he's become something of a national institution: a performance poet who became a regular on Terry Wogan's chat show in the mid-1980s; a stand-up comedian who appeared on the Labour Party's Red Wedge tour; a *Coronation Street* actor for more than a decade; presenter of the wildly successful BBC2 series *Robot Wars;* a BBC Radio 6 Music presenter and internationally successful funk DJ. Craig is still best known for his role of the slobbish, curry-guzzling technician Lister in the cult sci-fi sitcom *Red Dwarf.* He lives in Altrincham, Cheshire with his wife Jackie Fleming and their two daughters.

Craig on the decks at WOMAD music festival

"WHAT INTERESTS ME ABOUT GADGETS IS, DOES IT DO A JOB? IS IT REPLACING SOMETHING FOR THE BETTERMENT OF PEOPLE? I WANT TO KNOW IF ANYBODY CAN USE IT"

How do you see your role on *The Gadget Show*?
I suppose I'm the everyman figure. What interests me about gadgets is, does it do a job? Is it replacing something for the betterment of people? I want to know how much it costs, how long it will last, how easy it is to set up and use, and whether it replaces what came before it. I want to know if anybody can use it, or if it's just for people who are geeky. So I guess I'm very much the consumer on the show, asking questions people want to hear answered.

If the cast of *The Gadget Show* were a family, where do you fit in?
The way the show is structured, I'm like the headmaster and the rest of the cast have to show me their homework. Jon is the mad professor. Georgie is the girly swot who

always does her homework on time. Ortis is the crazy son, never off his phone. But, in reality, I'm also the naughty boy!

There's also the dynamic of you picking on Ortis…
Yeah, well that's not hard, is it? Ha ha! We're actually good friends, and it's all very good-natured. But, because of the format, you need someone to take the mickey out of. You can't take the mickey out of Jon, because he's almost a self-parody! But it works with Ortis.

Do you use a lot of gadgets at home?
Oh god, yeah. We love our gadgets and use them all the time around the house, although my wife has to set them up! We've got slow

CRAIG CHARLES

Debut on *The Gadget Show*:
10 March 2017

Home is: **Altrincham, Cheshire**

Go-to transport: **Camper van**

Did you know? **Craig is the only presenter who hasn't held a world record, but he did record two sessions for John Peel!**

cookers and air fryers and all the barbecue stuff. We've got the best Wi-Fi set-up and great televisions, with multi-room speakers throughout the house. I've even got a hot tub with Bluetooth speakers! And I've got an 18-year-old daughter and a 23-year-old daughter who both live at home and use an awful lot of tech. The house looks like a depot for tech products half the time!

But I'm not into tech for tech's sake. I can't imagine spending a quarter of a million quid on a pair of loudspeakers. I'm not one for using an app to turn the lights on. It is a fascinating area, IFTTT, or 'If This, Then That', technology explored by companies like IFTTT, Zapier, Integromat or Apple HomeKit. It's all about having tech that responds to another event in the real world.

"I'M FASCINATED BY DRONE TECHNOLOGY. I THINK WE'RE ALL GOING TO BE FLYING AROUND IN DRONE CARS NOW THEY'VE GOT ANTI-CRASH TECHNOLOGY"

So if you come into the house after a long night, you can turn your lights on and boil the kettle at the same time. But I'm not someone who wants to use an app to boil a kettle. I'd rather just press a button!

Do you delight in playing the everyman?
I want to make things simple, so everyone can understand it. Jon has an encyclopaedic knowledge of technological products – he is very good at remembering serial numbers and technical details. Ortis has medical training, so he understands health products. And Georgie is really good at analysing the cultural implications of tech. They're all really, really bright people, and I'm the simpleton. I've long realised that it's good to knock around with people who are smarter than you, because they make you look good!

What are your favourite gadgets?
I really like drones. I'm fascinated by drone technology. I think we're all going to be flying around in drone cars, especially now

Above: Playing Lister (centre) in the cult TV show *Red Dwarf*, 1988

Right: On the set of *Robot Wars*, which Craig started presenting in the late 1990s

they've got all this anti-crash technology. I appeared on a TV show called *Don't Rock the Boat* last year where I had to row from Land's End to John O'Groats. Some of the aerial shots they got using drones made it look epic. In the old days you'd have to hire a helicopter, a pilot and a film crew to do this. Now you can do it all with just a drone. And I love robots, like the stuff made by Boston Dynamics. The robot dog, the robot dancer, they're absolutely stunning… but also quite frightening. You worry if Boston Dynamics could build robot armies like that!

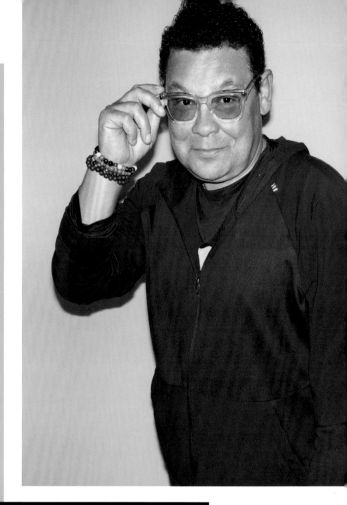

What music tech do you have?
I'm quite old school. I still prefer to use record players. When I DJ I use all sorts. I love vinyl, but it is very expensive and very heavy; you don't want to be dragging tons of records through the mud at festivals. Sometimes I'll use USB sticks, sometimes CDs are quite handy. You can cue a CD right on the beat, you don't have to do all this faffing around with dragging back a quarter turn for beat matching. Because my mixes are often quite intricate, it's nice to have that. And I'm not a scratcher, I'm a mixer, so I don't need vinyl a lot of the time. On radio, we rarely use moving objects in the studios. It's horrific to discover, but a lot of radio stations don't have a record player any more, or even a CD player. It's all digital!

> **"IT'S HORRIFIC TO DISCOVER, BUT A LOT OF RADIO STATIONS DON'T HAVE A RECORD OR CD PLAYER ANY MORE. IT'S ALL DIGITAL!"**

Below: Boston Dynamics' agile robot dog, Spot, which can capture data and cross rugged terrain

What cars do you own?
I've got a Mercedes and a 1966 Land Rover Series II, short wheelbase, the one with the headlights in the middle and the canvas back. But when I'm going to gigs I usually travel in my Mercedes Marco Polo camper van. It's all beautiful, with a fridge, a freezer, a cooker and a shower. An old mate of mine is my driver. It's the best way to travel, because I can get some sleep while on the move. It's all about time management when you're as busy as I am!

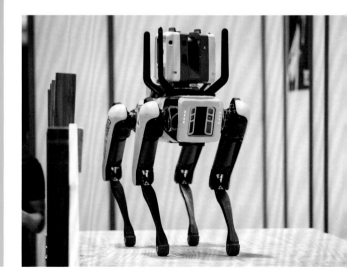

"WHEN I'M GOING TO GIGS I USUALLY TRAVEL IN MY MERCEDES MARCO POLO CAMPER VAN. AN OLD MATE OF MINE IS MY DRIVER. IT'S THE BEST WAY TO TRAVEL BECAUSE I CAN GET SOME SLEEP"

Right: Craig and Ortis channel their inner Jedi

Are you still a poet?
I've been writing these epic *Scary Fairy* poems that are 30 minutes long. I did them on Radio 2 with a great score, featuring the BBC Philharmonic Orchestra, and I'm now trying to turn them into stop-frame animations, made by Mackinnon & Saunders in Altrincham, the animators who did *Postman Pat* and *Bob the Builder*, as well as *Fantastic Mr Fox* with Wes Anderson and *Mars Attacks!* with Tim Burton. So that's my big project now!

Do you ever fancy reading poetry on *The Gadget Show*?
There was talk initially of me doing a little rhyme in each episode, like I used to do at the end of *Robot Wars*. But I always want to move on to new adventures. I did *Coronation Street* for 10 years, then I did *I'm a Celebrity Get Me Out of Here*, then my brother died while I was in the jungle and I had to leave. And that led me to re-examine my career a bit. *The Gadget Show* has been a wonderful adventure and, because of the nature of the show, it's always changing, always different, always new.

CRAIG'S PREDICTIONS

My first mobile phone was the Nokia 'Brick' in the late 1980s. Then I had another Nokia, the 'Mars Bar'. Then came the Ericsson flip-down one, where you looked like you were scouting for life forms on *Star Trek*.

Then the smartphone came in, where you couldn't change the battery, it was in the phone. Maybe that's where things started to get a bit dystopian!

But phone tech is amazing these days. They say there is more power in your mobile than they used to put a man on the moon. And, of course, the processing power increases all the time.

There's that observation called Moore's Law, where the number of transistors in an integrated circuit board tends to double every two years, so technology gets smaller and

smaller. The logical conclusion is that it will end up on a molecular level, so we will have a chip installed into our brains which will allow us to Google anything.

You won't need to learn anything, you'll just ask Google or Alexa. But you then have to worry about who runs Google! See, that's the thing with technology – it can always be used against you in the end!

MAGIC MOMENTS

The Gadget Show's fab four pick their favourite moments when filming the series... and reveal what was really going on behind the scenes

Craig on the decks

I was in my element when I got to test turntables in a record shop

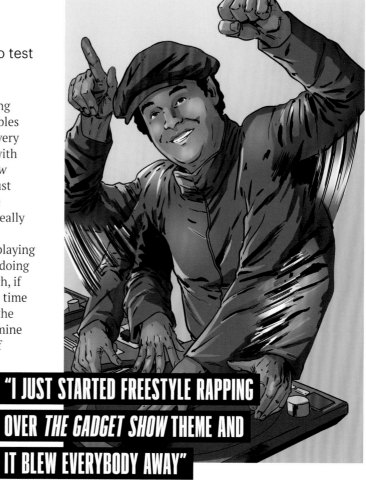

I am in my comfort zone talking about music tech and turntables and stuff like that. So, I was very happy to go test turntables with Jon. We made an edit of *The Gadget Show* theme and made it quite funky. Then I just started rapping over it, just doing some freestyle raps and stuff like that. And it really blew everybody away.

It was nice to feel as though I wasn't playing catch-up on the show for once. We were doing something that I knew about just as much, if not more than them, because a lot of the time these guys are so tech savvy. My role in the show is to be the common man, so I examine technology from the consumer's point of view. Is it easy to set up? Is it worth it? Is it tech for tech's sake? Does it improve our lot?

That's my role, which I'm quite happy to play. I think a lot of tech shows put people off gadgets and technology, simply because they make you feel ignorant. Jon is the one who can tell you the model number of every single device. I can't. But I do know that my CD decks are Pioneer CDJ-2000s and if I'm using turntables I'll use Technics. You can put vinyl though the Pioneers, too. It can handle

> ## "I JUST STARTED FREESTYLE RAPPING OVER *THE GADGET SHOW* THEME AND IT BLEW EVERYBODY AWAY"

vinyl, CD and USB. Most people these days put their record collection onto a USB stick, which is the size of a very small cigarette lighter.

It's nice to see vinyl and record players coming back. I've got a wall that's full of records. Even a couple of years ago, my kids would ask me what they were. I think people are starting to rue the fact that they threw their record collections away now, because vinyl has come back so strong.

We filmed the test in Bristol in a record shop that was closed because of Covid, but they opened it for us. And it was great because DJ-ing didn't happen in lockdown. I did three gigs in 18 months; I've had what you might call a kick in the wallet! Streaming sets isn't the same. I go off the vibe in the room. DJ-ing to a camera isn't the same; you may as well stick Spotify on and put it on shuffle.

Craig showed his DJ skills reviewing turntables with Jon

Ortis in flight

I felt like Iron Man with jet engines strapped to my arms

Richard Browning hands over the jet suit to Ortis

There was this massive barn in the middle of nowhere with a fighter jet outside, and we were told to start filming because Richard Browning, the main guy from Gravity Industries, would come to meet us. Then I heard the roar of jet engines and this humanoid form came flying out of the barn, over the jet, and then landed. That was a genuine surprise from me on TV.

He was wearing a leather suit and had mini jet engines strapped to him, one on his back and two on each arm. He was basically wearing a fuel tank. We talked about the technology for a bit and then Richard asked if I wanted to experience it myself. He explained that I wasn't allowed to fly because there wasn't enough time to teach me, but he let me strap on the jet engines and feel the power.

I braced myself. I could feel myself getting lighter, being lifted off the floor. I wasn't allowed to take off, but I experienced a form of weightlessness. The weight of the heavy jet engines on my arms disappeared. They kept saying, please make sure your arms are pointed downwards. Don't point your arms back otherwise you'll do a loop the loop.

I could feel my arms being elevated and when I battled against that and pointed them down, I could feel myself being raised onto my tiptoes. It was amazing. And it was loud, so loud. They're real jet engines.

I'm big into comics. Iron Man is one of my favourites in the MCU [Marvel cinematic universe] on screen. I think Robert Downey Jr really brought him to life. So yeah, it was full superhero geekery. Generally, *The Gadget Show*

"I COULD FEEL MYSELF BEING RAISED ONTO MY TIPTOES. IT WAS AMAZING. AND IT WAS LOUD. THEY'RE REAL JET ENGINES"

points me in a direction and says "do this" and I do it. But I have immense fun.

Another challenge that sticks in my mind is when I used flight simulators for six weeks before jumping into a real plane and flying, without any previous experience. I had a very nervous flight instructor, who did well to sit on her hands while I taxied, took off, flew a circuit and landed. The landing is obviously the most technical and terrifying part. She didn't touch her controls, she let me do all of it, but she was ashen. We touched down and I filled the air with jubilant expletives, because it was such a massive accomplishment, one of my biggest in television and in life.

Georgie in VR

I'll never forget the VR bungee jump the producers made me do twice

Georgie doing a real-life bungee jump wearing a VR headset

I t was one of the scariest things I've ever done. I'm not particularly scared of heights, but it was the first time I've ever done a bungee jump and it was in virtual reality (VR). The production company, Happy Finish, came up with this crazy plan for how to use the Oculus Go wireless VR headset and I was the guinea pig.

There's a great, simple game on the Oculus Quest 2 called *Richie's Plank Experience* where you have to walk the plank. It's fun but some people genuinely can't do it. I can do it quite easily; it doesn't scare me.

But this was real. I stepped off, with my headset on, and I was really falling. Then I hit the virtual ground and smashed through it into the flaming jaws of hell. I didn't know that was going to happen because they'd kept it under wraps for me. I was pumped full of adrenaline... but then the TV producers told me they wanted me to do it a second time, so they could film it without the headset. That was a shock.

When you're in VR, you can always tell yourself that it's just a game. But in real life your body knows that it's a real-life bungee jump and it would rather not step off the edge. The idea of doing it a second time was mentally draining, but I did it again... and they didn't bloody use it in the show, did they!

"I HIT THE VIRTUAL GROUND AND SMASHED THROUGH IT INTO THE FLAMING JAWS OF HELL. I DIDN'T KNOW THAT WAS GOING TO HAPPEN"

Still, I think I had a better time than the production team when we did the VR Guinness World Record in Westfield [read more about this on page 74]. I felt a bit sorry for the crew. I was in *Minecraft*, experiencing this amazing, colourful world that I was building, and they were just standing in a shopping mall.

I'm a big fan of VR. I have an Oculus Quest 2. You can only play it for a little bit because the headset gets a bit sweaty and heavy, but it gives a sense of scale and place that you can't really describe. One of my favourite things to do, if we have people over at Christmas or something, is give people their first taste of VR, to take their VR virginity! Friends take the headset off, flabbergasted by how amazing the experience is. They're like: "I feel like a million dollars!" It's just so much fun.

Jon on wheels

I drove my dream Alfa Romeo. I also had a food fight with Anthea Turner

ars are one of my greatest passions, so many of my favourite moments on *The Gadget Show* are on four wheels. When they wanted me to test podcasting tech, I got to drive one of my all-time favourite cars, the Alfa Romeo SZ, for the first time and make a podcast about it.

It was splendid. The car is a classic from 1989–91. I've had the T-shirt for ages with 'Il Mostro' on, which is its Italian nickname, so it was a joy to actually drive one. It has a wonderful engine and makes a very nice noise, so the sound was perfect for testing podcast recording equipment.

Working on *Top Gear* in the 1980s and 90s, I have a good memory of driving, say, a Lamborghini Countach or a Ferrari Testarossa, but this was a car I'd missed.

Old cars are often very disappointing though. One of my biggest disappointments in life was driving the Aston Martin DB4 Convertible from *The Italian Job*, which they didn't actually push off a cliff with a bulldozer; they used a different car. So, the car survived and I remember driving it after it had been restored and finding it quite underwhelming.

Why cars? It's a sensation of speed. To be in control of a machine is nice. It gives you tactile feedback in so many different ways. Psychologists have come to the conclusion

"THE ENGINE MAKES A VERY NICE NOISE, SO THE SOUND WAS PERFECT FOR TESTING PODCAST RECORDING EQUIPMENT"

that cars are good for you. People are generally happier when they have cars and are able to get about under their own volition.

The other great thing about *The Gadget Show* has been meeting so many interesting people, who I wouldn't have got to meet otherwise. I've tested kitchen gadgets with the Hairy Bikers, I've been rowing with Dame Kelly Holmes, tried set-top boxes with Noddy Holder and walking boots with Brian Blessed.

The funniest was with Anthea Turner. I was testing washing machines and the director had the idea that Anthea and I should have a food fight both dressed in white. We'd throw food at each other three times and then see which washing machine did the best job. Unfortunately, Anthea's agent intervened at the last moment, so I had to be the one on the receiving end of all the foodstuffs. I was covered in food and food colourings; she wasn't. The washing machines did a brilliant job of cleaning up the results, actually.

Star guest Anthea post food fight, pre washing machine test

WOULD LIKE TO A

YOU PLAY GAME?

VIDEO GAMES ARE MAINSTREAM THANKS TO AFFORDABLE TITLES, NETFLIX-STYLE SUBSCRIPTION SERVICES AND GAMES YOU CAN PLAY ON YOUR SMART SPEAKER. IT MEANS WE'RE ALL GAMERS NOW, SAYS **JORDAN ERICA WEBBER** OF *THE GADGET SHOW*

PRESS START

My whole life I've been playing games and they've changed. They used to be kind of hidden away. Fewer people played them, and those who did had to buy an expensive machine, read magazines to find out what was new, and kick family members off the TV or hunch over the PC in the corner of the sitting room to play. But now, with a playful device in almost every pocket, and each new generation making gaming more mainstream, games are everywhere. And I'm delighted.

Sure, there are still people who spend a fortune on the latest gear, like the elusive Nvidia GeForce RTX 3080 graphics card, and the latest Sony PlayStation 5 and Microsoft Xbox Series X consoles. The games with the biggest marketing budgets are usually the same kinds of stories – they often involve a lot of guns. And there are still whole subcultures, growing bigger by the day, which value skill above all else, like the enormous esports scene. I can see why so many people tell me that they don't think of themselves as gamers.

But as the gaming world has grown, it has also diversified. Some of the biggest games in the world right now – like *Among Us*, *Roblox* and *Fortnite* – are playable with your pals across a whole range of devices, cheaply or even free. We've all become more used to hanging out with our loved ones online and games like these are great social spaces, too.

Speaking of affordable games, there's a whole world of unique little independently developed games out there, downloadable from online stores such as itch.io and Steam. I recommend the gorgeous adventure game *A Short Hike*, the coming-of-age visual novel *Butterfly Soup* and the minimalist island-building game *Tiny Islands* by David King.

The ways we find games to play have changed, too. Microsoft has made a big push for Netflix-style subscription services with Xbox Game Pass, which gives you access to a library of more than 100 games you can play on an Xbox console, Windows 10 PC, or even (via the magic of the cloud) an Android device, for a monthly fee. With big hitters like the blockbuster *Halo Infinite*, and more individual titles like the dungeon-crawling dating sim *Boyfriend Dungeon* and the open-

Above: Jordan, expert gamer at *The Gadget Show*

WE'VE ALL BECOME MORE USED TO HANGING OUT WITH OUR LOVED ONES ONLINE AND GAMES LIKE *AMONG US* AND *ROBLOX* ARE GREAT SOCIAL SPACES

Left: Esport fans at the 2019 *League of Legends* World Championship final

IT HAS BECOME INCREDIBLY POPULAR TO PARTICIPATE IN GAMING CULTURE WITHOUT EVEN LAYING YOUR HANDS ON A CONTROLLER

world exploration game *Sable*, there's something for everybody.

Smaller names are trying subscription-style models, too: Playdate is a cute little yellow handheld console, with a crank controller as well as buttons, from *Untitled Goose Game* publisher Panic, with industrial design by Teenage Engineering. It comes with a 12-week 'season' of games, with two new titles delivered wirelessly each week. Some of the best indie developers in the world have made games for this unique device, like *Crankin's Time Travel Adventure*, *Echoic Memory* and the playfully named *Snak*.

Of course, it has become incredibly popular to participate in gaming culture without even laying your hands on a controller, thanks to live-streaming services like Twitch. Swing by my channel (twitch.tv/jericawebber) to say hi and see what all the fuss is about. Naturally, I stream a lot of brand-new games and *The Sims 4*, but you can find people playing basically anything, from the free-to-play battle royale *Call of Duty: Warzone* to the multiplayer piracy

Below: Panic's handheld console Playdate gets new games delivered via Wi-Fi

play date

GET YOUR GAME ON

FROM THE BEST-SELLING TO THE PRETTY MUCH ICONIC, HERE'S *THE GADGET SHOW*'S PICK OF TOP GAMES TO ENTERTAIN, EDUCATE OR TURN YOU INTO A HARDCORE FAN

Among Us The scrappy little multiplayer game that had us betraying our friends in space and spawned countless memes. Sus.

Roblox A phenomenally successful game creation system and online platform, full of weird and wonderful virtual places to hang out.

Butterfly Soup A heartwarming visual novel about queer young women of colour playing baseball, playable for free on itch.io

A Short Hike A gorgeous, award-winning indie game about climbing a mountain and helping the people you meet along the way.

Pokémon Go The augmented reality (AR) creature-catching mobile game played by millions worldwide that grossed more than $6 billion in four years.

Lost Your Marbles The marble-rolling choice-driven narrative adventure from Sweet Baby Inc, one of the first games playable on Playdate.

Sea of Thieves A colourful multiplayer pirate adventure for Windows and Xbox, developed by beloved studio Rare in the Midlands.

Call of Duty: Warzone One of the best-selling video game franchises worldwide delivers its take on the free-to-play battle royale.

Ratchet & Clank: Rift Apart The latest iteration of the popular platform adventure from Insomniac Games, exclusive to the Sony PlayStation 5.

Fortnite The free-to-play battle royale game that made Epic Games its billions, played by basically every teenager everywhere.

Soup Pot A cooking game that lets you experiment in the kitchen risk free, complete with real-world recipes to drool over.

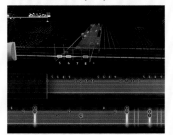

Rocksmith+ The guitar game that promises to actually teach you to play guitar, providing real-time feedback as you play along.

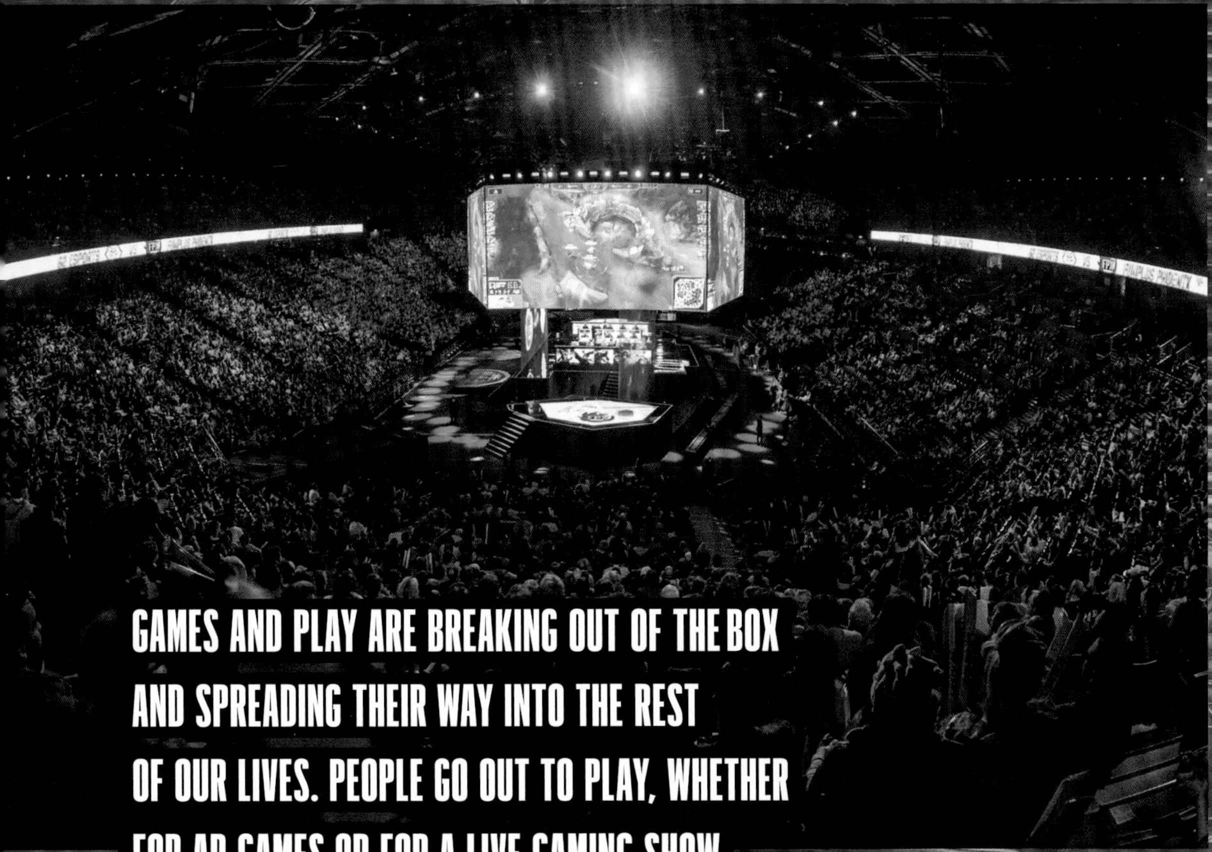

GAMES AND PLAY ARE BREAKING OUT OF THE BOX AND SPREADING THEIR WAY INTO THE REST OF OUR LIVES. PEOPLE GO OUT TO PLAY, WHETHER FOR AR GAMES OR FOR A LIVE GAMING SHOW

adventure *Sea of Thieves*. With storylines and 3D graphics comparable to Hollywood blockbusters, it's no wonder more and more people are joining the audience. Twitch started with gaming live streams but has since expanded into other areas – you can also tune in to watch people providing commentary on football, making music in their studio or creating visual art.

Parents need not worry about their kids' interest in games getting in the way of future career ambitions. Even if they don't make it as an esports pro, YouTuber or Twitch streamer, there are plenty of game-related careers these days, not least in actually making the things. And you don't have to learn complex programming languages. *Game Builder Garage*, for the Nintendo Switch, uses a colourful drag-and-drop interface and funny writing to teach you the basics of level design.

Above: The *League of Legends* esport event, held in an arena and viewed by millions

Right: Games such as *Zombies Run!* make exercise fun

If you want to get a bit more advanced, try the PlayStation game creation system *Dreams*, which people have used to make some truly stunning games and art. Even the hugely popular *Roblox* is less game itself and more a platform through which to find games other people have made using its tools.

Games and play are breaking out of the box and spreading their way into the rest of our lives. People go out to play, whether that's for augmented reality (AR) games like *Pokémon Go* (or developer Niantic's new *Pikmin*-themed game) or for a live gaming show like *WiFi Wars* or *The Incredible Playable Show*.

Playing at home doesn't have to mean sticking to a screen, either. Got one of those smart assistants on the kitchen counter? You can get Google to play *Twenty Questions* with you, ask Amazon's Alexa for a game of *Mystery Sounds*, even dive into an audio-only representation of the incredibly popular fantasy role-playing game *Skyrim*.

People are even learning through play. Ubisoft's *Rocksmith+* teaches you bass and guitar by detecting how well you can play along to songs. *Soup Pot*, from independent developer Chikon Club, is a cooking game with real recipes. And geographic discovery game *GeoGuessr*, which plops you down in a random place on Google Street View and asks you to guess where you are, has introduced an educational version.

Of course, this kind of gamification has long been used to encourage people to exercise. I use the mobile audio game *Zombies, Run!* to motivate me on my morning jog. And I've played a lot of *Ring Fit Adventure* on my Nintendo Switch. You could even shell out for the Playpulse One, an exercise bike with integrated video games. Or make your friends laugh by using the GPS tracking on a running app like Strava to draw silly pictures.

Everyone plays games of some kind and I mean *everyone*. That means creating more diverse characters, so more people get to see heroes who look like them. And it means

I USE THE MOBILE AUDIO GAME *ZOMBIES, RUN!* TO MOTIVATE ME ON MY MORNING JOG AND I'VE PLAYED A LOT OF *RING FIT ADVENTURE*

making games more accessible to people with different needs. Games like *Ratchet & Clank: Rift Apart* have a range of accessibility options that open the game up to players with visual impairments or those who have difficulty pressing buttons at speed. Microsoft has gone the extra mile and created the Xbox Adaptive Controller (see page 59), which lets players with limited mobility customise their game controller to their own specific needs.

Being playful is human nature and technology simply enables it. So throw away your preconceptions about what a video-game player is like. We're an overwhelmingly welcoming bunch. If you want more tips on where to jump in, just keep watching the show. I'll be sure to let you know!

GAME OVER

Jordan puts
Craig's gaming
skills to the test

PLAY LIKE A PRO

WITH MILLIONS OF DOLLARS UP FOR GRABS,
PROFESSIONAL ESPORTS PLAYERS INVEST IN
THE LATEST AND GREATEST TECH, AS ESPORT
HOST FRANKIE WARD HELPS EXPLAIN

Esports tournaments – competitions between especially skilled players of multiplayer games – can rack up millions of viewers on live streaming platforms. As professional esports host Frankie Ward (pictured) points out, it's very similar to watching traditional sports: "As supporters, you share the common goal of wanting to see your favourite teams and players succeed. You unite over your love of the game and you come away having shared an experience with thousands of people in the same space."

As with traditional sports, you don't have to be a skilled gamer to get a kick out of spectating. But with prize pools of millions of dollars up for grabs, it's no surprise that more and more kids want to get into esports. And those who do want to become professionals need professional gear. "A monitor with a fast frame rate is essential," says Frankie. "Most players will have a 240Hz monitor as standard, such as the BenQ ZOWIE XL2540/XL2546." The monitor also has built-in shields at both sides to block out distractions.

Even headphones are chosen carefully. They have to be noise-cancelling, says Frankie, "to stop players from hearing commentators in the arena or hints from the audience. Tournament organisers such as ESL currently use the Bose QC range, playing white noise as an extra layer of disguise."

When choosing a mouse, again speed is of the essence, but latency is so low these days that a lot of players are going wireless. Frankie says Logitech is the market leader here: "Its G502 Lightspeed is already considered a design classic many top gamers would swear by." And keyboards tend to be the clicky mechanical kind, with popular manufacturers like Razer designing with esports athletes in mind for its Huntsman Tournament Edition, and Ducky offering smaller "tenkeyless" options (a compact keyboard without a number pad), like the Ducky One 2 TKL for portability.

From top: BenQ Zowie XL2546K monitor; Razer Huntsman Tournament Edition keyboard; Logitech G502 Lightspeed mouse; Bose QC 35 II headphones

Toy story

The best hi-tech toys and games for kids aged one to... 100. Just think of it as the ultimate gift guide for others and yourself

STELLINA

VAONIS STELLINA

This 80mm aperture refractor telescope can be set up in minutes and controlled via an app. Choose a celestial object and the telescope will track it. Stellina also filters out light pollution and stacks images to create digital photos in real time.

GOCUBE EDGE

This smart Rubik's Speed Cube comes with Bluetooth, so the companion app knows precisely which tiles are where at all times. It can teach you to solve the cube and set challenges. Battle it out with friends or strangers.

RAZER KRAKEN BT KITTY EDITION

Razer makes stunning games PCs and peripherals. This wireless headset features kitty ears that can be programmed to light up in colours, synched with your other Razer Chroma kit. It has low latency and high kawaii.

AIRSELFIE AIR PIX

Think of this not as a palm-sized drone but as a camera that you can gently throw into the air for a selfie. It flies back and frames a photo of your group, returns and sends it to an app for sharing on social media.

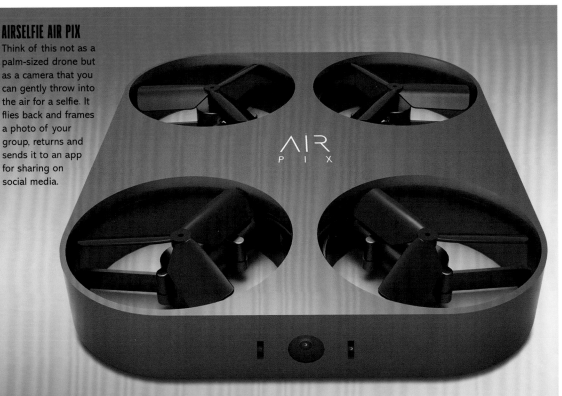

GIGA LOUNGER

An inflatable that literally won't let you down. At a click, the built-in pump inflates the lounger in 60 seconds. Everyone will want a go. You can also plug your phone into the pump's USB to recharge it while you relax.

LOOG GUITAR

A cute learner guitar for youngsters. Flashcards and an app teach basic chord triads. The smallest version is barely bigger than a ukulele; its three strings match the first three on a larger guitar, making it easy for kids to graduate to six. Amp and speaker are built in.

TEENAGE ENGINEERING POCKET OPERATOR

An affordable mini synthesiser and sequencer that comes in lots of crazy editions for music making. Teenage Engineering has loaded them with samples of everything from office equipment to vintage Capcom video games.

ANYCUBIC VYPER

Looking for a first 3D printer? This is affordable, faster than its predecessors and, crucially, has a self-levelling bed, a must for newbies. The build volume is respectable and it can print in four types of plastic (PLA, TPU, ABS, PETG) as well as in wood fibre.

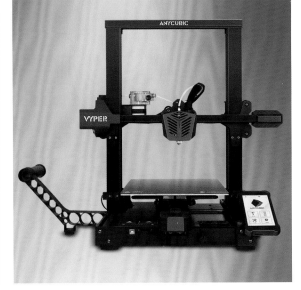

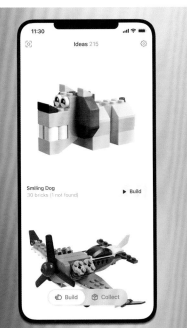

BRICKIT

Spread out your Lego bricks then use this app to see what you can build. It takes a photo and scans them, giving you instructions and showing you where each brick is in the pile. Possibly the best app idea ever.

XBOX ADAPTIVE CONTROLLER

Designed in partnership with the gaming and disability community, this controller is the size of an A4 pad of paper, with large, programmable buttons. Third-party assistive devices can be plugged in, too, opening up Xbox and PC gaming to everyone.

Cathedral of books

This Chinese grand design is bliss for bibliophiles. If only all bookshops were this magical

It's a bookshop, but not as you know it. This vast room is part of a 973-square-metre Zhongshuge bookstore in Dujiangyan City in south-west China's Sichuan province. The store was designed by X+Living, a female-led Shanghai practice. Lead architect Li Xiang had designed other Zhongshuge bookshops before, all of which delight in tricking the eye. Look up and you'll see a mirrored ceiling, giving the impression the room reaches to the sky. There's no giant ladder; the shelves within reach are full of books while digital printing decorates the higher shelves with 'book pattern film' designed to inspire. Arches allow views into other spaces, like the children's whimsical reading area featuring colourful mushrooms and shelving in the shape of pandas and bamboo. A work of fantasy.

Sounds elegant

Beauty is in the ear of the beholder. Classic hi-fi separates have been updated with the latest tech for music-lovers of all generations

1 ECLIPSE TD307MK3
Eclipse monitors are loved by musicians, producers and sound engineers for their high fidelity. These squeeze the loudspeaker tech into a smaller shell. The egg shape ensures that no two internal radii are the same, so sound isn't coloured by internal reflections and resonances.

2 ASTELL&KERN A&FUTURA SE180
A&K specialises in premium digital music players – think lossless high-res files, exquisite materials and four-figure price tags. This portable model adds interchangeable DACs (digital-to-analogue converters) to tailor your sound.

3 MELCO N100
This shoebox-sized digital jukebox stores 5TB (5,000 gigabytes) of high-res digital audio, a lifetime of music. It supports streaming anywhere in the home. Import your own music or buy and download studio masters for the best possible quality.

4 ELIPSON CHROMA 400 RIAA BT
This French-made turntable boasts a glossy lacquered finish and all mod cons. It has a light, rigid carbon tonearm, a built-in phono preamplifier and an aptX HD Bluetooth transmitter for wirelessly streaming your vinyl.

5 MCINTOSH MHA200
Valve (vacuum tube) amplifiers are known for their warm analogue sound. This modern valve headphone amp delivers the very best high-fidelity personal listening from any analogue source. Plus, the valves glow beautifully.

6 GRADO STATEMENT SERIES GS3000E
Wired hi-fi headphones hand-built in Brooklyn. They're crafted in Cocobolo wood, from Central America, used to make instruments because of its rich, deep and impactful musical character. Oversized 50mm signature drivers ensure a large soundstage.

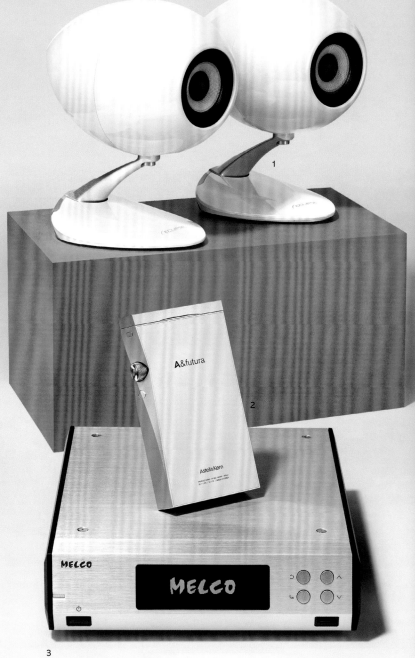

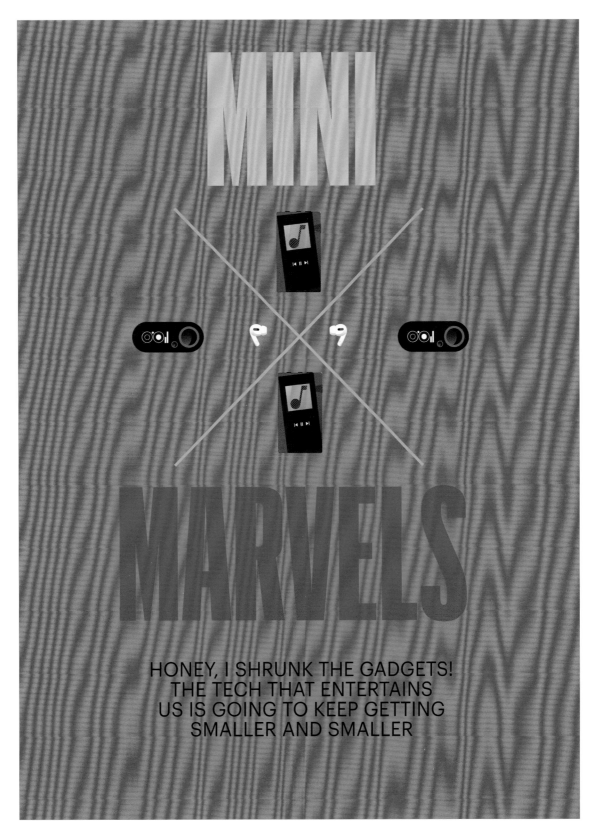

MINI

MARVELS

HONEY, I SHRUNK THE GADGETS!
THE TECH THAT ENTERTAINS
US IS GOING TO KEEP GETTING
SMALLER AND SMALLER

The year 1977 was seminal. The nation celebrated The Queen's Silver Jubilee, the Sex Pistols released their only studio album – and the personal computer was born. At the beginning of January, Apple Computer was incorporated; in the same month, the world's first all-in-one home computer, the Commodore PET, was demonstrated at the Consumer Electronics Show; and in August, Tandy announced its TRS-80 Model 1.

Computers, once room-filling beasts, were coming home. Moore's Law, named after one of Intel's co-founders, had predicted it back in 1965. The law says that the number of transistors on microchips doubles every two years. As computing power grows exponentially, so our gadgets shrink. Today, a computer far more powerful than 1977's museum pieces can fit into a matchbox thanks to the Raspberry Pi Zero.

More than 40 years ago, Microsoft co-founder Bill Gates had a vision that there would be "a computer on every desk and in every home". We're already there in developed countries. That is, if you count the powerful computers in our games consoles, set-top boxes and phones. There are very few homes with none.

Computing power lies at the heart of next-gen games consoles from Microsoft and Sony, but notably Nintendo has gone much smaller with its Switch console, designed to be played on the move as well as at home. It's smaller by design but you can plug the original Switch into a TV at home for the best of both worlds.

The early plasma TVs of the 1990s were 'flat' but surprisingly deep compared with the wafer-thin screens of today. OLED (organic light-emitting diode) technology means pixels are self-lit. This does away with backlighting, so screens can be just a couple of millimetres thick, which enables even affordable TVs to be much slimmer than their predecessors. It makes incredible designs possible, like the LG Signature OLED R, a pricey 65-inch screen that simply rolls away to be hidden in furniture when not in use.

Meanwhile, projector technology has shrunk so much you can enjoy a cinema-style big screen projected onto your wall from a surprisingly small gadget. The BenQ GS2 is a cute 14cm cube and the Prima pico projector is barely bigger than a smartphone.

Hi-fi aficionados are still in love with big speakers, separate systems and the analogue authenticity of vinyl on a great turntable. But these have met their match with the advent of high-resolution audio files that boast higher sampling rates than CDs. The higher the sampling rate, the harder it is to differentiate a digital audio source from analogue, even from a small portable player. Team a premium portable like the Astell&Kern A&norma SR25 with top-notch headphones and even audiophiles would be hard-pushed to hear the difference.

You can also upgrade your digital music by plugging a premium DAC (digital-to-analogue converter) between your digital source and your hi-fi. This bypasses your device's basic DAC and puts it through a superior chip which extracts even more detail from your digital music.

PROJECTOR TECHNOLOGY HAS SHRUNK SO MUCH YOU CAN ENJOY A CINEMA-STYLE BIG SCREEN PROJECTED FROM A SMALL GADGET

Above left: The Raspberry Pi, a credit-card sized computer used for anything from learning to code to robotics

Above right: BenQ GS2, a 14cm cube wireless LED projector, comes with built-in TV apps and a Bluetooth speaker

WALLOP'S TOP TIP

Consumer advice from *The Gadget Show's* Harry Wallop

EASY E-READING

"The BorrowBox app lets you borrow a digital edition of a book your local library has in stock. Otherwise, head to the Google Play Store, Amazon or Kobo for a large selection of free ebooks."

The Arcam Black Box, first made in 1988, was the size of a hi-fi separate and the first such standalone DAC. Now you can buy portable hi-fi DACs the size of a pack of cards, like the Chord Mojo and AudioQuest DragonFly. There are also small DACs specifically aimed at gamers, like the SteelSeries GameDAC. Inch for inch, they're the best sonic upgrade you can buy.

Large over-ear headphones physically shut out the rest of the world, especially if you go for a closed-back design, but the latest tiny wireless earbuds give them a run for their money. The Sony WF-1000XM4 earbuds use a wireless version of high-resolution audio and active noise cancellation to block out the sound of the outside world.

The fact that you can store thousands of songs digitally on your phone, or stream millions from online services, deserves kudos, too. Your music collection is practically infinite yet fits in your pocket. The same is true for an entire library of electronic books, though it's easier on the eye to read them on the paper-like display of ebook readers like the Amazon Kindle Oasis.

Last but not least: tablets and phones. They boast incredible power but the screen size means they can't get much smaller. Or can they? New, folding OLED screens let your devices shrink further. The Samsung Galaxy Z Fold3 5G is the size of a standard smartphone but it opens like a book to reveal a tablet-style screen that's twice the size. And the Galaxy Z Flip3 5G phone folds in half when not in use to become supremely pocketable. Voice control functions and paired devices like smart watches also mean your tiny phone can increasingly stay in your pocket.

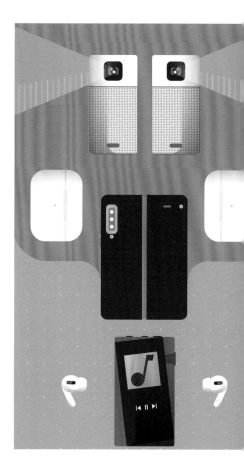

ORTIS DELEY ON HIS FAVOURITE SMALLS

"My favourite small bit of tech is my smart watch, although I resisted it for quite a while; I know how I feel during a workout, so I rarely used my Fitbit. But I switched from Android to Apple recently and that move was led by my desire to have an Apple Watch.

It means I use my phone less. I've limited the amount of information that comes to the watch to just SMS messaging, emails and some business alerts. I can see those at a glance, so I'm less likely to get drawn into looking at Instagram or Twitter.

I think the ideal bit of shrinking tech at the moment is earbuds. I haven't worn big headphones for more than a year. I have Beats earbuds on my iPhone and Samsung Galaxy Buds on my Android phone, which I still carry with me."

MUSIC ON THE MOVE

FROM HANDBAG-STYLE RADIOS TO ICONIC WALKMANS VIA THE HEAVYWEIGHT BOOMBOX, THE DEVICES WE PLAY OUR MUSIC ON HAVE JUST GOT MORE AND MORE PORTABLE

PHONOGRAPH 1877
Thomas Edison created a hand-cranked machine that would play back sound recorded on fragile tin foil wrapped around a grooved cylinder.

GRAMOPHONE 1887
Emile Berliner patented the wind-up forerunner of the modern record player, but it only became popular when vinyl discs arrived in the 1930s.

ROBERTS RADIO 1932
The iconic British portable radio was originally the size of a suitcase, but its R66 portable valve radio (1956) and transistor-based RT1 (1959) are the inspiration for today's retro designs.

BOOMBOX LATE 1970S
Portable music came to the street... if you didn't mind balancing more than 10kg on your shoulder. 1981 models like the JVC RC-M90 and the Sharp GF-777 helped give rise to hip-hop culture.

SONY WALKMAN 1979
The portable cassette player heralded the first time we could plug in headphones anywhere and tune out the rest of the world. You'll never forget your first mixtape.

SONY DISCMAN 1984
CDs brought superior digital music quality to music on the move, however early models could skip if you ran for the bus.

MP3 PLAYER 1997
Remember Napster? The music industry was terrified that music-lovers would share illegally downloaded files, but ultimately most preferred to pay for the convenience of streaming services.

APPLE IPOD 2001
Before the iPhone was the iPod, which put your iTunes music collection in your pocket. Its intuitive user interface, iconic design and colourful advertising made it a hit.

ASTELL&KERN 2013
This premium digital audio player with superior DACs (digital-to-analogue converters) plays high-resolution digital audio for the ultimate in music fidelity on the move.

The entertainers

Watch and listen in style with the best home entertainment technology, for indoors and out

FLEXOUND PULSE

Sound and its natural vibration are built into this chair, turning home cinema into a 4D-like sensory experience and making video games more immersive. You feel the sound, even frequencies that are too low to hear.

SAMSUNG THE TERRACE
A weatherproof smart TV, designed to be permanently mounted outdoors, with an anti-reflection screen and vivid QLED 4K HDR picture. Great for outdoor movie nights, barbecues and keeping the kids entertained.

NEBULA SOLAR PORTABLE
A rechargeable battery with three hours' playback time makes this your go-anywhere projector. It delivers a 1080p HD HDR big picture with 400 lumens for brightness whether indoors or out. Android is built in for streaming apps.

SONY LSPX-S3
This wireless speaker looks like a candle as it lights up and flickers. The glass tweeter radiates sound in all directions, while Sony's LDAC file format supports high-res audio streaming over Bluetooth. Buy singly or pair for stereo sound.

NOVETO SOUNDBEAMER 1.0

An infrared 3D camera precisely tracks your ears to give you personal sound. Two arrays, each with 128 tiny speaker drivers, direct narrow ultrasonic beams to your ears, where they interact to create little bubbles of sound only you can hear.

LG ÉCLAIR

A small soundbar that's shaped more like a lozenge. Despite its size, it squeezes in upwards-firing drivers to deliver cinematic 3D sound with Dolby Atmos and DTS:X. It has 3.1.2 channels including the separate subwoofer.

URBANISTA LOS ANGELES

The world's first self-charging wireless headphones, thanks to a solar panel on the headband. Switch between advanced hybrid active noise cancelling and ambient sound mode to stay aware of your surroundings.

IFI HIP-DAC
Add a superior digital-
to-analogue converter
to improve sound quality
from all digital sources.
This portable one looks
like a hip flask and has
a built-in headphone
amp. Use it on the move
or at home.

GEORGIE BARRAT

THE THINKER

GEORGIE LOVES A DRIVERLESS CAR
AS MUCH AS THE NEXT NERD, BUT
ALSO LIKES GETTING TO GRIPS WITH
TECH'S HEFTIER ISSUES, FROM ITS
CULTURAL IMPACT TO PACKING
A POLITICAL PUNCH

eorgie Barrat graduated with a first in English literature from King's College London, studied radio production, then got into broadcasting. She worked on London Live, *ITV News*, Aled Jones's ITV show *Weekend*, Channel 5's *The Wright Stuff*, ITV's *Tonight* and Channel 5's *The Saturday Show*, before landing a job as presenter on *The Gadget Show* in January 2017. She lives in south London with her husband.

How did you get into tech?

Technology was the common language in my household. I am the middle of three kids. Both my brothers are big gamers and, even before I started playing games myself, I would watch my big brother play *Zelda* and read out the instructions to help him get to the next stage. My younger brother is very much Gen Z. He isn't particularly big on social media but I would say all aspects of his life take place over the net. I've always been fascinated by how they both interact with technology.

If *The Gadget Show* were a family, how do you see it?

Craig and Jon are the parents. Ortis and I are the schoolchildren. Ortis thinks we're like stepbrother and stepsister, which I find hilarious. And Craig takes the mickey out of Ortis, so you've got this dysfunctional parent-child relationship. Craig is always very apologetic about that, by the way: "Sorry Ortis, mate, but this is in the script. It's a bit brutal!" But we all get on really well.

When did you start writing about tech?

I had a tech podcast at university. I don't even think the word podcast was used then, but it was a slot that I made for the student radio station, something that could be downloaded or listened to online. It was one of the many pre-recorded shows that you'd put on the server and maybe five listeners would give it a whirl.

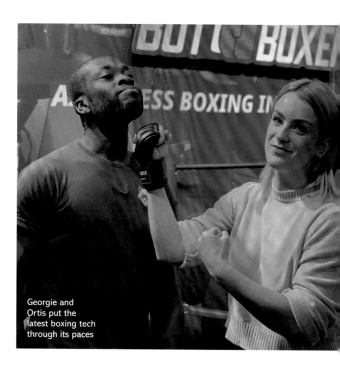

Georgie and Ortis put the latest boxing tech through its paces

"VIRTUAL REALITY IS BRILLIANT FOR GIVING YOU A SENSE OF FREEDOM AND SCALE. IT WOULD BE GREAT ON A LONG-HAUL FLIGHT"

When I became a broadcast journalist, I started specialising more in technology. I got interested in how tech was shaping us as a society, about the impact of social media on our everyday lives, about how apps are changing the way we live. This raises massive questions: What is the future of autonomous vehicles? What impact will robotics have on the job market? How will the next iteration of Alexa impact us? I started thinking critically and looking at how all these technologies are genuinely – and this is no exaggeration – transforming every part of how we live and interact. The amazing thing is, all this has happened within the life of *The Gadget Show*.

Is it true you have a world record?

I got the world record for being on virtual reality (VR) for 26.5 hours, playing

GEORGIE BARRAT

Debut on *The Gadget Show*:
10 March 2017

Home is: South London

Go-to transport: Train

Did you know? Georgie holds the world record for time spent in virtual reality (26.5 hours)

Above and left: Georgie during her marathon VR session, which won her the Guinness World Record

Minecraft. Was it fun? I wouldn't use the word fun, it was long and slightly gruelling, but it was fine. The Guinness World Records people allowed me a five-minute toilet break every three hours or so. Some people get motion sickness in VR, which would be horrendous. Luckily, I didn't. I think the producer was a bit annoyed – they wanted me to suffer! In the game you do a lot of virtual jumping around. So, when you take the headset off, you feel very static and slow. You realise you're in Westfield in Stratford at 4am and you think, "Oh god, this is drab, maybe I'll put the headset back on!"

Are you a big VR fan?
I think it's brilliant for giving you a sense of freedom and scale. It would be great on a long-haul flight. It's really developed over the life of *The Gadget Show*. When I started, the headsets were half decent but they were still tethered to a computer. The HTC headset cost more than a grand. Then came the consumer version of the Oculus Rift in 2016. You didn't have to connect it to a computer, all the hardware was stored in the headset itself. But it's still got a few more iterations to go. Play VR for more than 15 minutes and it's hot, it's heavy, it's uncomfortable. If someone else has been wearing the same headset, you're like, ew!

What other products have evolved during the life of *The Gadget Show*?
Augmented reality (AR) is another example. That was something really niche. Then *Pokémon Go* came out and there was a huge explosion. Now people are using it every day when they put a filter on their phone, adding bunny ears to a Snapchat or Instagram photo. It's become humdrum, to the point that people aren't even aware that it's a very sophisticated use of tech.

You studied English literature at King's – do you bring a critical theory edge to technology?
Definitely. It provides a field day for cultural analysis. I'm fascinated by the effect technology is having on humanity. Everyone

is on their devices for multiple hours every day. The population is connected in ways that have never happened before. This has huge effects, culturally and politically, for better and worse. We've seen how big data influenced elections in 2016 – both the Brexit referendum and Donald Trump winning the presidency – and how targeted advertising on social media could have a huge effect on democracy.

Technology has also been an incredible tool to help us work remotely during the pandemic. But it has taken us millions of years to evolve as social creatures and we still crave social connection. Creative people are still going to have their best ideas through face-to-face conversation; colleagues will get more done over coffee than on Zoom.

What happened to your planned TED talk on feminism and technology?
It didn't happen, for various reasons, but I would love to deliver it one day. It was about how technology is liberating women and facilitating the next evolution of feminism. Tech can help get women back into the workplace, help with parenting, tap into a bigger community and share ideas. We can use big data to analyse the points we need to improve regarding gender inequality and other institutional inequalities. There are exciting possibilities and potentially terrible drawbacks. Although tech can connect women and fuel the next wave of feminism, women receive a hell of a lot of online abuse. It can give with one hand and take with another. It's a complex picture but, ultimately, it can help us move forward.

Have gadgets had a progressive effect?
Throughout history, when you look at things that were traditionally the work of women, gadgets like the washing machine have liberated women from household chores. So, what's the next iteration of that? What traditionally holds women back from returning to work after childbirth? You could use AI to help run a household, organise

"THE POPULATION IS CONNECTED IN WAYS THAT HAVE NEVER HAPPENED BEFORE. THIS HAS HUGE EFFECTS, CULTURALLY AND POLITICALLY"

your diaries. You could use autonomous vehicles to liberate people from carting their children around the place. Technology can also spread ideas, democratise education, spread transparency and so on.

Tell us about your podcast, *Sleep Life*…
It's a podcast that I host with the writer Alex Goldstein, about one of my favourite subjects: sleeping! We spend a third of our life asleep but it's something we rarely talk about or understand. Sleep is the foundation of everything we do while we're awake. So, we talk to experts about how the right kind of

> **"SOCIAL MEDIA PLATFORMS EARN MONEY BY HAVING YOUR EYES ON THEIR PRODUCT, SO THEIR ONLY INCENTIVE IS FOR YOU TO BE ON THAT PRODUCT FOR AS LONG AS POSSIBLE"**

sleep can positively impact your creativity and your productivity; about how to deal with sleep when travelling; about how sleep works when different people in relationships and families have different sleeping patterns. We even talk to athletes about how sleep can enhance sporting performance.

How are you with 'tech hygiene', for instance, keeping your phone away from the dinner table or the bedroom?
It's something we all need to be aware of. Being sucked down a YouTube wormhole is not a productive use of your time. A lot of social media platforms earn money by having your eyes on their product, so their only incentive is for you to be on that product for as long as possible. These things are designed to suck you in and keep you there. We're now reaching a turning point where people are thinking: hmm, I don't want to be spending so much time on this device. You need to break that umbilical cord with your phone!

Above: Hosting her *Sleep Life* podcast with Alex Goldstein

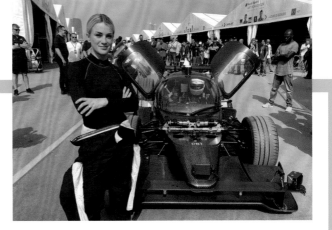

GEORGIE ON DRIVERLESS CARS

We did a special in Hong Kong to cover Roborace, where autonomous electric vehicles [pictured] compete, although eventually they'll race without drivers – it will just be a battle of coders. I was strapped in and it was driving better and faster than me, changing gear and everything, which was terrifying and incredibly alien.

Cut to a year or two later, in Las Vegas. We booked a Lyft autonomous car as a taxi. It turned up with a person behind the wheel, for legal reasons and because it was still a prototype, but the person didn't intervene at all. Within minutes, being driven by an autonomous vehicle turns into the most natural thing in the world. You quickly put your trust

in it. It's like being escorted by a very good, confident driver.
I live in London so I don't have a car, but can see there won't be many car purchases in me before I get an autonomous vehicle. No matter how much you think technology pushes us outside our comfort zone, if tech does the job better than humans, we will quickly hand over that job.

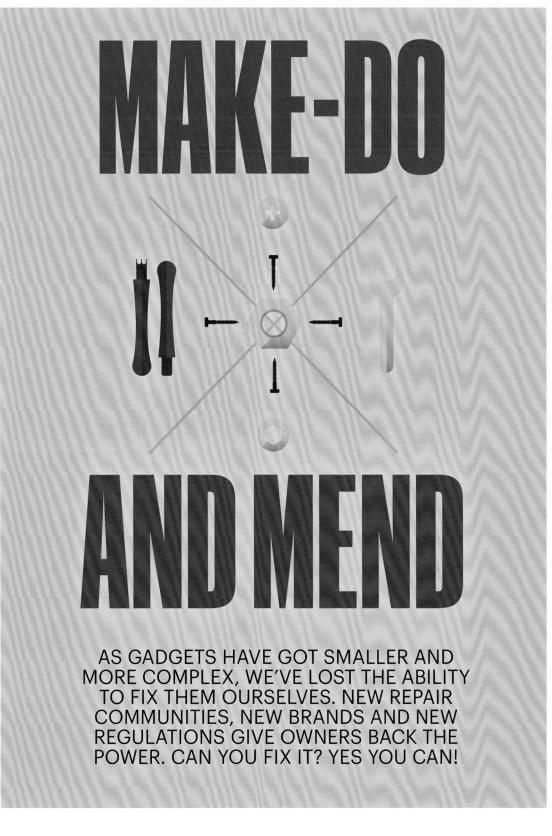

MAKE-DO
AND MEND

AS GADGETS HAVE GOT SMALLER AND
MORE COMPLEX, WE'VE LOST THE ABILITY
TO FIX THEM OURSELVES. NEW REPAIR
COMMUNITIES, NEW BRANDS AND NEW
REGULATIONS GIVE OWNERS BACK THE
POWER. CAN YOU FIX IT? YES YOU CAN!

The moment you drop your phone, your heart sinks. Will the screen be cracked? Will the phone even work? Are your photos backed up on the cloud or not? And why is it that you own this device but are powerless to repair it when it goes wrong?

One of the well-known wartime posters urged Britons to "mend and make-do to save buying new". That thrifty repair culture continued long after the Second World War. Most of us grew up with a parent, or at least someone in the family, who could fix the car as well as darn a sock. Electricals lasted for years, and when they broke they were repaired.

Decades of miniaturisation later and we have e-waste mountains and phones that only survive a year or two. Even cars are harder to repair now. Council amenity sites are piled high with white goods. You can spend hundreds on a new appliance, but if it breaks down after a year and a day and you don't have an extended warranty, you're on your own.

A new internet community is springing up to empower us to repair our own tech. The biggest site, iFixit, boasts hundreds of thousands of solutions. If your gadget maker has used proprietary fasteners and kept its repair manuals secret, help is at hand.

The site combines community with commercial: you can find free repair guides or buy a kit that gives you precisely the parts

> **IF YOUR GADGET MAKER HAS USED PROPRIETARY FASTENERS AND KEPT ITS REPAIR MANUALS SECRET, HELP IS AT HAND**

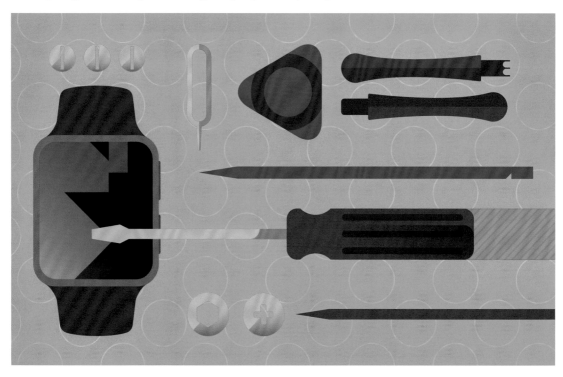

Left: Dualit's fully repairable Classic toaster is a kitchen companion for life

and specialist tools for the job. These include tiny screwdriver bits for proprietary fasteners, suction cups to lift smartphone screens and the gloriously named 'spudger' for prising tech open.

If you're repairing a phone, for example, you typically do need to remove the screen to get inside, even for seemingly simple repairs like replacing a battery that no longer lasts a day. You'll also find yourself reaching for a hair dryer to warm the adhesive that holds the screen on. Generally, the pricier the gadget the more it's worth repairing. iFixit sells kits for repairs to popular phones, laptops, tablets, even the Apple Watch and Nintendo Switch games console.

If you'd prefer tech repair lessons in person, search for a local 'repair café'. These community initiatives are springing up around the country with the aim of sharing knowledge, lending specialist tools and cutting e-waste.

The result is truly green: cutting carbon emissions is about more than looking for a gadget that's energy efficient. Keep your gadgets alive for longer and you don't just save money, you save the materials and energy that go into making and transporting a replacement.

KEEP YOUR GADGETS ALIVE FOR LONGER AND YOU DON'T JUST SAVE MONEY, YOU SAVE MATERIALS AND ENERGY

The government is finally playing its part, too. New 'right to repair' rules came out in 2021. These force manufacturers to publish repair manuals and make replacement parts available for at least a decade. But the rules only protect your right to repair specific kitchen appliances and TVs; smartphones and laptops aren't covered.

Built-in obsolescence sees tech need replacing far too often. New brands are responding to consumer demand for an alternative. Companies like Fairphone and Framework want you to keep your gadgets for many years, empowering you to upgrade them as you go.

Fairphone is proud of its ethical supply chain. It sources materials responsibly, uses conflict-free minerals and champions workers' rights. Its phones are modular and designed to be easy to repair or upgrade at home. Fasteners are standard screws. You still need a steady hand, but you don't need specialist tools.

Framework offers a similar ethos for laptops. Assemble them yourself or buy them pre-assembled, then replace modules any time to repair or upgrade.

Some traditional electrical brands have always championed repairability, like Dualit toasters. Buy a Classic model and, if something stops working, you can get the relevant spare part. The design is simple, so you'll only need an element or a timer because there's nothing else to go wrong. Similarly, Magimix food processors are workhorses for which you can buy all manner of spare parts, even second-hand ones on eBay. Like a car breaker's yard, there's money to be made by selling used parts rather than scrapping them.

FAIRPHONE IS PROUD OF ITS ETHICAL SUPPLY CHAIN. IT SOURCES MATERIALS RESPONSIBLY AND USES CONFLICT-FREE MINERALS

Other brands and retailers are confident enough of the longevity of their products that they offer free, extended guarantees for peace of mind. British brand Mitchell & Brown has a seven-year warranty on its TVs, while John Lewis offers five-year guarantees, and at least two years on laptops and large kitchen appliances, and three years on its own-brand electricals. Join Richer Sounds' free VIP Club for a six-year guarantee on most TVs, projectors and wireless multi-room audio systems. With many of these guarantees the gadgets might be replaced, rather than repaired, but it will still save you money.

JON THE GADGET FIXER

The Gadget Show's JON BENTLEY on the pleasure of repairing old tech

As a child I used to buy broken tellies and repair them to boost my pocket money. I could identify a dodgy capacitor, a dodgy valve or a loose connection. Then I'd advertise them in the paper or sell them to teachers.

Repairing technology has got a lot harder, thanks to microelectronics, but I'm always prepared to have a go. I managed to keep a dishwasher going for 24 years. I re-soldered the electronics, I replaced the control board, I replaced the bit that the spray arms attach to… but when I couldn't find the source of the leak underneath, I decided that was enough.

If you like the idea of repairing older technology, the British Vintage Wireless Society has meetings. You can find suppliers who take an old capacitor that's leaking and put a new one inside, keeping the old canister. You can even upgrade an old wireless to add Bluetooth!

WALLOP'S TOP TIP

Consumer advice from
The Gadget Show's
Harry Wallop

USING IFIXIT

"If you've got the patience and the skill with a screwdriver, it's well worth considering repairing larger consumer electronics such as laptops and games consoles. The savings are substantial."

And finally, the reply to the most common tech repair question: should I put my wet phone in uncooked rice? The answer is no! There's no evidence that putting it in rice absorbs the moisture. Also, it's easy for grains of rice to get stuck in the headphone socket, if your phone has one.

Instead, you should turn the phone off immediately, resisting the urge to check if it still works, because an electricity-plus-water short circuit kills phones. Rinse quickly if the dunk was in anything other than clean water. Dry off the surface water but don't use heat. If your phone's back cover comes off and you can take out the battery, do it fast. Remove SIM cards and memory cards, too. Keep all the covers off. Then dry your phone for at least 48 hours. The best way to do this is with a shop-bought drying bag like the AF Tech-Rescue kit, otherwise improvise one with a sealed container and lots of those little sachets of silica gel desiccant that arrive with new purchases. Save them and store them in an airtight container ready for phone-in-the-loo emergencies.

THE MOST COMMON TECH REPAIR QUESTION: SHOULD I PUT MY WET PHONE IN UNCOOKED RICE? THE ANSWER IS NO!

PERFORMANCE ENHANCERS

Get the competitive edge and recover faster
with the latest hi-tech equipment – essential kit
that works for these top British sports stars

When Roger Bannister became the first man to run a mile in under four minutes, his one concession to technology was wearing shoes he had commissioned from cobbler GT Law & Son that weighed just 128g. They were 50 per cent lighter than the shoes he normally ran in. On the day of the record attempt, he was sharpening their steel spikes on a grinder in St Mary's Hospital, when one of his fellow students said,

"You don't think that's going to make much difference, do you?"

Today's athletes share Bannister's determination. From Premiership footballers and racing drivers to heavyweight boxers and Olympic snowboarders, if there is a marginal gain to be made from a hi-tech device, they take it. And if those gadgets can boost their performance, they can certainly help yours. Take their advice for upping your game and that 'Couch to 5k' will be a breeze…

ANTHONY JOSHUA OBE
World heavyweight boxing champion and Olympic gold medallist

"The most important thing I have learned from my training, as both an amateur with Team GB and in the professional ranks, is that rest and recovery is key. That's why I don't go anywhere without my Pulseroll Mini Massage Gun [pictured]. It's a massage device that uses vibrations to flush away lactic acid and increase blood flow deep into my muscles, helping to reduce any tightness.

Using the Mini Massage Gun before and after training helps me bring the required intensity to the gym day after day, meaning I can maximise every session during a training camp. I've been using it, and other devices like vibrating foam rollers, as part of my training programme for almost four years and have noticed a significant difference. It's great for relieving those niggling muscle knots

and also for relaxing when I'm at home. The other useful thing is that the Mini Massage Gun is the size of my phone, so it's really easy to carry around. I can use it when I am travelling (it has a built-in rechargeable battery) or whenever I feel any pain or tension."

JAMIE CHADWICK
Professional racing driver

"My Whoop Strap [pictured] is the one piece of fitness tech that I would recommend to anyone. It's the only wearable fitness product I've ever had that I don't really notice I'm wearing. Because it is so light and comfortable, I can wear it day-to-day and even while I'm racing. Like me, it performs on the track and off-road – so it monitors my physical responses while I am competing and gives me a score to indicate how hard I have been performing. The results are so accurate that I have even heard of one athlete who realised they had caught Covid after noticing a change in their data.

The charging is also really easy. It has a charger you slip on, so you never have to take the device off, which means it is constantly producing data. It's based around your day 'strain' and your HRV (heart rate variability) recovery, as opposed to just the fitness activity you have done, which is great and gives you a much broader set of results.

It's something I have become heavily reliant on, from first thing in the morning to see how I have slept, to charting my recovery after training or racing."

TOMMY FLEETWOOD
Professional golfer

"I know some players don't like having a watch on when they play, but I love wearing my TAG Heuer Connected Golf Edition watch [pictured], both on and off the course. The features are well thought-out, making the feedback on my game much more precise. And, of course, it looks cool.

One of the elements I really like is the 'distance' shot feature. It means I can see how my drive is performing on the golf course, factoring in the actual conditions I am playing in. It has been especially useful when I've been testing my new clubs because it tells me just how far I am hitting the ball. I also love the fact that they have added the scoring feature. Being able to keep track of your scores on your wrist without an old-fashioned pencil and a scorecard is a very cool feature.

It has been exciting to be involved in the future of the app and constantly talking about how we can make it better for users and golfers worldwide. To be honest, I can't think of a golfer whose game wouldn't benefit from using it."

JENSON BUTTON MBE

Former F1 world champion

"Although my Formula 1 racing days are behind me, I still drive competitively and go karting when I can, so I need to be fit. Luckily, I've always enjoyed training, and these days I regularly go road biking and trail running. Recovery is also a big part of my fitness routine, so to avoid overdoing things I love riding my ebike.

My VanMoof S3 [pictured] looks brilliant. It has a simple, understated design and certainly doesn't scream 'electric bike'. The gearing is automatic and it has a 504Wh battery, which pulls the front wheel along quickly and effortlessly. I can also lock and unlock it with a simple app on my phone. For me, it has been a complete game-changer for everyday bike use.

Since having two kids, and with three dogs, I inevitably have to buy a lot more than I used to when popping to the shops. Cycling home when I've got a fully laden backpack, that Turbo Boost button is perfect for giving me a little kick up the hills. Plus, it's the bike version of KITT from that old TV show *Knight Rider*. So, what's not to like?"

KATIE ORMEROD

Olympic Snowboarder

"I can tell you honestly, my favourite bit of tech that I use when snowboarding has to be my Therm-ic heated socks [pictured]. As I compete and train in temperatures well below zero, I often get very cold feet, so wearing these socks enables me to stay comfortable on the slopes and perform at my best in all weather conditions, especially as the heat is directed throughout the whole of each foot.

The battery clips in at the top of the sock which sits just below my knee – so it's well out of the way of my snowboard boots – and is unnoticeable under my outerwear, which is important to me. They have different levels of heat that can be altered by pressing the button on the battery. Having the convenience of a USB makes the socks easy to charge and they are great for travelling as they come with a compact case.

The socks have a long battery life, so I don't have to worry about getting cold feet during the day. The charge lasts throughout the duration of my training sessions and competitions."

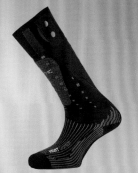

DOM PARSONS

**Former skeleton racer and
Olympic bronze medallist**

"Fractions of a second can be critical
in skeleton racing. Once we push the
sled off the block at the start, we only
have gravity to help us accelerate
down the track to speeds of up to
130km per hour. Aerodynamic drag
reduction, control and handling, runner
efficiency and bespoke athlete fit on
the ice are all key factors that affect
performance. Much like a race car,
there are a whole range of setup
changes that we can make to our sleds
to suit different tracks and weather
conditions – and most importantly
to suit our preferences as athletes.

Bromley Sports continually pushes
the boundaries in all these areas,
producing some of fastest skeleton
sleds in the world. I started using
a Bromley skeleton sled [pictured]
in 2014 and the options in setup
and adaptability of the sled helped
me to make a huge step forward in
my sliding, so that I could compete
with the best in the world."

JORDANNE WHILEY MBE

Grand Slam tennis champion and double Paralympic medallist

"I use the Pulseroll Electric Foam Roller Pro [pictured] on a daily basis as part of my training routine and, since I've started using it, I've noticed a significant improvement in my recovery. I spend 15 minutes a day on the roller and it helps to prevent the dreaded DOMS (delayed onset muscle soreness) and increase flexibility. The roller I use comes with five different speed intensities controlled by a small remote – ideal for my warm-ups and cool-downs.

I travel a lot to Grand Slams and tournaments and the roller has now become a key part of my kit bag – it's nice and lightweight to carry around, too. Not only is it scientifically proven to improve circulation, blood flow and recovery from DOMS by 22 per cent, it also increases range of movement (in other words, flexibility) by 14 per cent.

The best part is that due to the vibrations there is almost no need to roll. The research and the work Pulseroll has done with the NHS also gives me the added confidence in the product and the benefits of percussion massage therapy."

MARK CAVENDISH MBE
Professional road and track cyclist

"I have worn Oakleys throughout my career and in my opinion the latest Oakley Kato glasses [pictured] are one of its best ever. I love the unique mask-like design and in particular the science behind it, which improves a rider's field of vision due to the oversized frameless lens that wraps right around the nose, sitting perfectly on the face. On a personal level, that really helps when you are sprinting in a pack and you have rivals to your left and right.

Oakley's Prizm lenses are the other element I love. This technology enhances colours, depth perception and contrast so you can see more detail in terms of your surroundings and contours of the road, making your vision even better than what you would be able to see with the naked eye. Believe me, when you are travelling at high speeds, it's imperative to have total clarity of vision when you approach the finish line!"

MARCUS RASHFORD MBE

**Manchester United and
England footballer**

"When it comes to training and playing at the highest level, the one bit of tech that gives me the edge are my Nike Mercurial Superfly 8 Elite boots [pictured]. I'm a speedy player and this boot helps support my natural pace. It fits like a natural extension of the foot and this helps me maintain control at high speeds. The traction enables quick acceleration and braking in all directions, integral for the way in which I play across the forward line.

The standout tech is the upper made of Vaporposite, a grippy grid mesh with premium lining, and the Nike Aerotrak plate, which aids responsiveness. They come in bright colours and lightweight materials – and are apparently inspired by the natural geometry of a dragonfly's wings to drive maximum efficiency with minimal weight. The level of thought that goes into these boots is insane, but I love how all these little elements come together. Most importantly, the boots are comfortable and reliable.

The Mercurial range is constantly evolving with the game. It's one of Nike's longest-serving boots, clocking in at over 23 years now, and it's helped the athletes I look up to, such as the OG Cristiano Ronaldo himself and many more. If that boot can develop their game, it can help mine, too."

Wearables

Clever clothes,
new performance
fabrics and hi-tech
accessories

VOLLEBAK SOLAR CHARGED JACKET

A thin phosphorescent
membrane absorbs
light and re-emits it
slowly, for hours. Cyclists
and runners will find
the bright green glow
protective. Geeks will
go into a dark room
and write on themselves
with a torch.

ALTBERG DALESWAY WALKING BOOTS

This traditional Yorkshire bootmaker has moved with the times. The Dalesway is made from Lorica, a vegan microporous alternative to leather, backed with a Sympatex breathable, waterproof lining, and has a Vibram sole.

GOT BAG ROLLTOP BACKPACK

Made from 100-per-cent ocean plastics, this waterproof bag is a world first. It uses 3.5kg of PET, recovered by fishermen on the shores of Indonesia, which is shredded and turned into a polyester yarn. It comes in four colours.

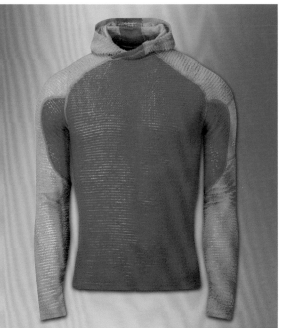

OMM CORE HOODIE

A ridiculously lightweight layer (115g) thanks to insulating fabric PrimaLoft Active's open structure which traps air for warmth when stationary but maximises airflow on the move. Read: you won't keep taking it on and off.

BUFF THERMONET HINGED BALACLAVA

For winter sports or winter generally. PrimaLoft thermal insulation ensures the balaclava is warm but breathable. It offers full-face cover but has a hinged opening – handy for a snowboarding selfie.

VOLLEBAK GARBAGE WATCH

This prototype watch is built purely from electrical waste. Each element is repurposed from rubbish and Vollebak will share the story as it sources the components. Get yourself on the waiting list – it's open now.

RAB MICROLIGHT ALPINE JACKET

The Microlight jacket has had an eco facelift. Its down filling is recycled from sleeping bags or bedding, which is cleaned and coated in Nikwax's hydrophobic treatment. The inner and outer performance fabrics are recycled, too.

ARTILECT DARK HORSE LEGGINGS

Artilect's zoned leggings are woven from Nuyarn, which is made by spinning merino wool along a high-performance filament. This results in a better natural fibre that's fast drying, with more stretch and loft than regular merino.

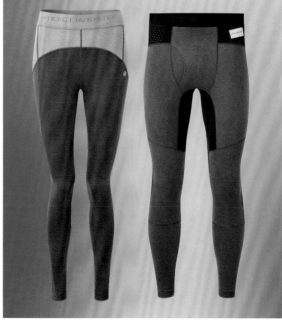

DIVOOM PIXOO BAG
Programme your own
emoji or pattern on this
backpack's 13-inch, 256
LED pixel display via an app.
You can even add a wireless
remote control to your bike
handlebars and use the
display as an indicator.

Your future could be in cyber

Fancy yourself as an action man or action woman? Technology can repair and even augment our bodies. Anyone can become bionic like Colonel Steve Austin – you don't even need six million dollars

CAMERA EYE
Rob Spence is a filmmaker who lost one eye in a shotgun accident as a child. Over the years, he has created several wireless video camera eyes to replace it, including this glowing red *Terminator* version. He's now working with a start-up to make 3D-printed prosthetic eyes, with the potential for built-in cameras.

HI-TECH PROSTHETICS
Double amputee and gamer James Young, nicknamed Metal Gear Man, worked with Japanese gaming giant Konami and the Alternative Limb Project to develop a bionic arm. Tiny muscle movements in James's back control the 3D-printed hand. The arm features a USB phone charger, smart watch, torch and mini drone.

THIRD THUMB

Designer Dani Clode created this 3D-printed thumb, positioned below the little finger. Its intuitive, wireless controls are two pressure sensors operated by your big toes. Researchers at UCL used the thumb to see how our brain adapts to augmentation, learning one-handed tasks like stirring a coffee while holding it with the same hand.

NERVE CHIPS

Self-experimenter Professor Kevin Warwick is perhaps the world's first cyborg. In 1998, he had an RFID chip embedded under his skin for smart home control. His work has since included neural implants to control a robot arm on another continent via the internet. He has also connected his nervous system with his wife's.

BLOOD SUGAR SENSOR

Some people with diabetes have to monitor glucose levels and manage them with insulin. A constant glucose monitor, like the Dexcom G6 (pictured), sits just under the skin and a transmitter sends data to smart devices wirelessly. Alerts warn of potential hyper or hypoglycaemia (high or low blood sugar) and can work with an insulin pump.

AUGMENTED BONES

Osseointegration uses a titanium rod implanted into the bone, so that an artificial limb can attach directly, rather than fitting a socket onto a stump. The implant has special pores so that the bone tissue has a strong adhesion. The aim is not only to avoid rejection, but also to integrate the human body and the artificial.

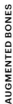

Get fit kit

Gadgets aren't just for entertainment. Tech can help us to be fit and healthy, whether with competitive sport, activity tracking or a fresh smoothie on the go

1 LIFESTRAW GO
This isn't just a sports bottle that filters out limescale, this is your go-everywhere, insulated survival bottle. Its membrane microfilter and porous, activated carbon filter remove bacteria, parasites and chemicals from the water.

2 GOPRO HERO9
The latest, greatest action camera for capturing all your adventures. It now boasts a front display, better battery life, 10-metre waterproofing, live streaming and voice control. Take 4K, even 5K video, and 20-megapixel stills.

3 SALOMON INDEX.01
A road-running shoe that's light and totally recyclable. It's made from two materials: a recycled polyester upper and a TPU foam. Both are recycled by Salomon at the end of the shoe's life and used in new footwear and fabric.

4 APPLE WATCH 6
Health and sports tracking are Apple Watch's biggest selling points, alongside its good looks. Track blood oxygen levels, take an ECG, or monitor sporting, sleep and activity. Team it with Apple Fitness+ for world-class workouts.

5 ANDEN DOMINO
This beautiful stand works with the wireless charging puck that's supplied with your Apple Watch. Made from raw brass and sustainable timber, it's the most elegant way to charge your watch bedside.

6 LEKI MCT 12 VARIO CARBON
These ultra-light carbon poles fold to 42cm in seconds and are designed for speed. They're ideal for cross-trail mountain walking, with a breathable mesh strap and sweat-absorbing, ergonomic foam grip.

7 JULBO EVAD-1
These futuristic smart sunglasses for sports such as running and cycling have a built-in heads-up display (HUD), putting your choice of live performance stats right before your eyes, and weigh just 35g. The photochromic lenses adjust rapidly to the light conditions.

8 POC MENINX RS MIPS
A lightweight but highly protective winter sports helmet, with a buckle that's easy to use when wearing gloves. Safety features include a RECCO reflector for avalanche rescue, NFC medical ID to store your critical information, and MIPS technology that reduces rotational forces on your head from an impact.

9 POC ZONULA CLARITY
The oversized, toric lenses, developed by POC with German optics experts Carl Zeiss, give these goggles an extra-wide field of view, and they're easy to swap for clarity in all conditions. The frame is made from 47 per cent bio-based materials to reduce its environmental impact.

10 NUTRIBULLET GO
A rechargeable personal blender, so you can make fruit smoothies or protein shakes anywhere. You can load it with fresh fruit in the morning but hit the button later in the day. One charge is enough for 20 uses.

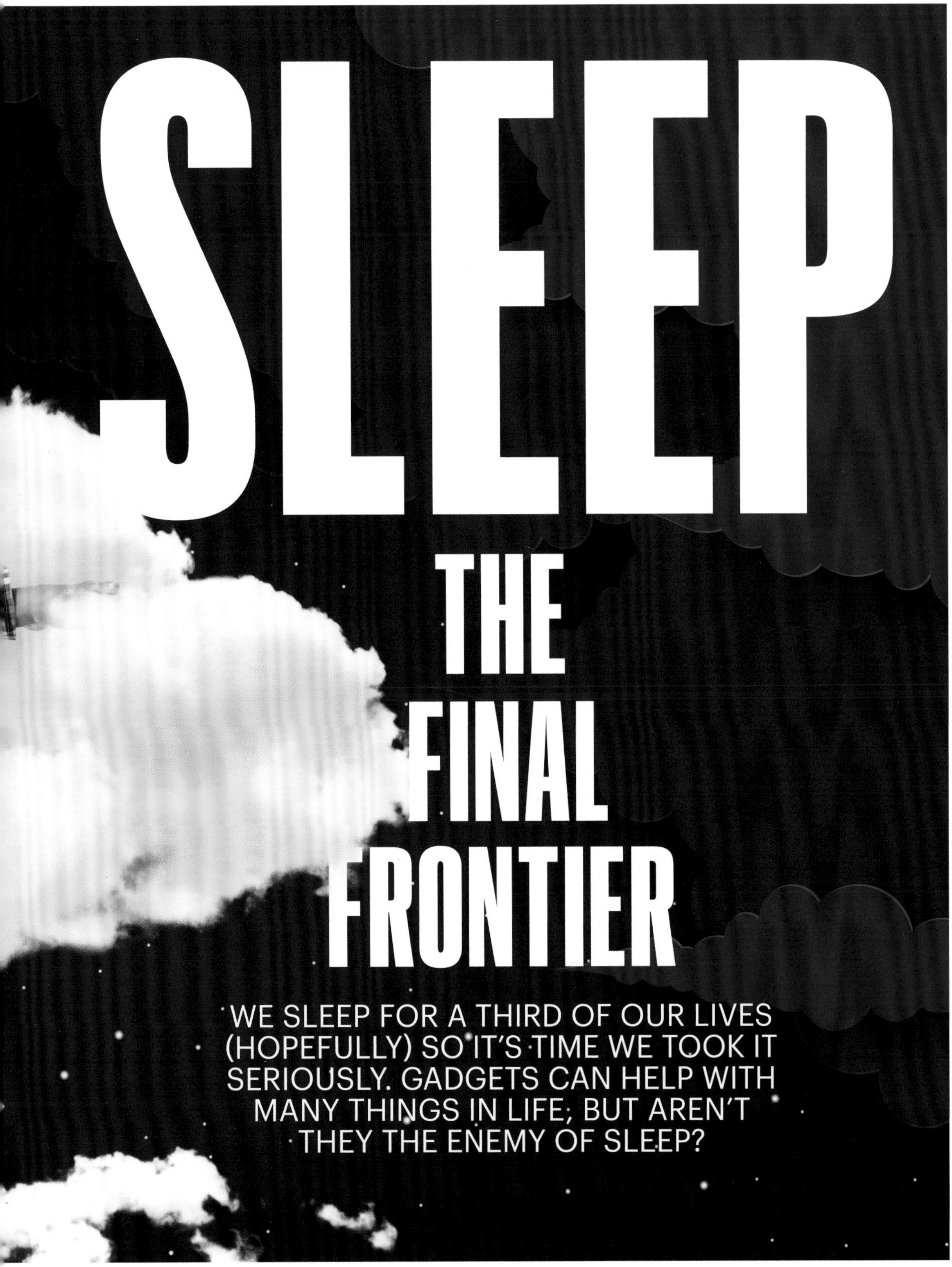

SLEEP

THE
FINAL
FRONTIER

WE SLEEP FOR A THIRD OF OUR LIVES (HOPEFULLY) SO IT'S TIME WE TOOK IT SERIOUSLY. GADGETS CAN HELP WITH MANY THINGS IN LIFE, BUT AREN'T THEY THE ENEMY OF SLEEP?

echnology isn't known for helping us get shut-eye. Sitting in bed having 'one last look' at social media and then doomscrolling for an hour is the enemy of sleep. Similarly, if you wake and immediately check your notifications, your day's priorities can get eclipsed before you've even started.

Screen protectors for phones, and glasses for gamers, promise to filter out the overdose of blue light that comes from our screens, potentially disrupting sleep. Putting your tech down a bit before bedtime, however, is a better plan. Or at least switching to different gadgets.

Voice assistants like Alexa, Siri and Google come into their own as you approach bedtime. Ask your smart speaker the time, the headlines, to pick music or a podcast, or have it read out your messages. It's less distracting than a screen. Or swap your screen for the eye-friendly E Ink display of ebook readers such as the Amazon Kindle Oasis and enjoy a book at bedtime.

SMART SPEAKERS WITH VOICE ASSISTANTS LIKE ALEXA, SIRI AND GOOGLE COME INTO THEIR OWN AS YOU APPROACH BEDTIME

Audiobooks and podcasts can be soothing and apps like Calm and Headspace (see overpage) curate content that's specifically designed to be soporific. Experiment and you'll find out whether you're someone who needs a story, a meditation or rainforest noises to fall asleep to.

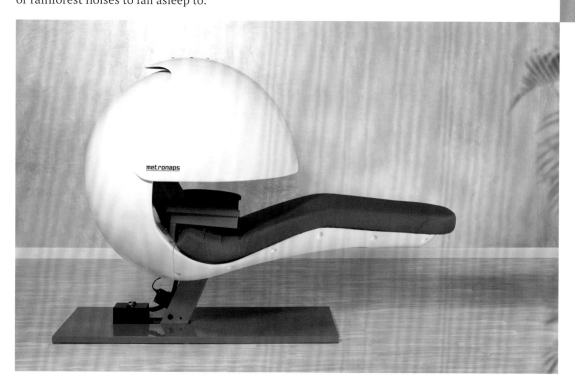

White noise is known to help with sleep (see Craig's tips for nodding off, overpage) and gadgets such as the Dreamegg D11 play a selection of sounds, from fan noise to ocean waves to a vacuum cleaner. The portable sound machine's design is safe for cots but thankfully doesn't look babyish. As an adult, it beats going to bed with Ewan the Dream Sheep.

Headphones are a great use of technology, whether we want to be lulled to sleep without disturbing a partner or hope to catch forty winks on a plane or train. The low-level white noise of active noise cancellation is good for travel but for bed, where comfort is king, look for small, well-fitting earbuds.

There are earbud-like devices created specifically for sleep. Bose Sleepbuds II look like tiny earbuds but they won't stream

Above: Ultra-slim Kokoon Nightbuds use personalised audio and monitor your sleep

Top: Sleeping pods in a capsule hotel, a popular concept in Japan

Opposite: Take a power nap in a MetroNaps EnergyPod, as installed in Google offices

your music. Instead, they're designed to block audible distractions and play your choice from a menu of nature sounds, tranquil soundtracks or noise-masking soundscapes in the Bose companion app.

Kokoon Nightbuds are a quarter of the thickness of many earbuds but have a band that goes behind your head. They don't just play sounds to lull you to sleep; they monitor you while you're sleeping and learn what works best for you. For example, whether to switch off or play noise-masking sounds to stop your sleep getting disturbed. Unlike the Bose, you can also fall asleep to your own choice of music or an audiobook.

Other smart devices that check in on you while you snooze at night include Fitbit or smart watches such as Apple Watch. Graphs in their apps quantify each night and give you weekly sleep averages that will either reassure you or explain exactly why you feel so tired.

As well as offering smart home controls, the second-generation Google Nest Hub is designed to sit next to the bedside and monitor your sounds and motion while you sleep. See your stats on screen the next morning or just ask, "Hey Google, how did I sleep last night?"

Google is well known for allowing its employees to have some shut-eye while in the office (as long as they get their work done when they're awake). The tech giant has even installed stylish MetroNaps EnergyPod sleep capsules in its US and UK headquarters, so that employees can take power naps.

How you wake up is just as important. Smart watches can wake you during light sleep, within a chosen time window. Lights like the Philips Somneo HF3651 use the gradual orange-to-yellow glow of a virtual sunrise to wake you in the morning.

> **GOOGLE NEST HUB IS DESIGNED TO MONITOR YOUR SOUNDS AND MOTION WHILE YOU SLEEP. SEE YOUR STATS ON THE SCREEN IN THE MORNING**

SLEEP? THERE'S AN APP FOR THAT

CALM

The British genius behind Moshi Monsters took a deep, mindful breath and then co-founded what's now the number one app for sleep, meditation and relaxation. Calm is full of curated content, from bedtime stories read by big names to sleepy soundscapes. Although you have to subscribe to the app, some of the content is free to try.

HEADSPACE

The co-founder of Headspace, another Brit, quit university to become a Buddhist monk. He now teaches and writes books on meditation and mindfulness. The app offers this and more, with music, sleepcasts and practical advice, plus tools for waking up and focusing. There's a free trial to see if it works for you.

CRAIG AND HIS FANS IN THE NIGHT

"I used to fall asleep to the sound of a hair dryer in the bedroom. I got the idea from Chris Gascoyne, who plays the part of Peter Barlow in *Coronation Street*. He gets through about six hair dryers a year!

I started doing it. But I've moved on because my wife now has two fans by the bed, one either side, and the noise of the fans has replaced the hair dryer for that kind of background *shhh*. I find it really, really relaxing for some strange reason."

Left: Google Nest Hub controls your environment as well as checking your sleep patterns

Far left: Bose Sleepbuds II offer a choice of sounds to snooze to

GEORGIE'S SWEET DREAMS ARE MADE OF THIS

" I co-host a podcast called *Sleep Life*. I've always been fascinated by sleep and well-being. My top tip is, don't charge your phone by your bed. It's important to have a little bit of space from our devices just before sleep and, crucially, first thing in the morning. I charge mine in the next room. But you can still have some tech in the bedroom. I have a Kindle and an Alexa Echo Dot by the bed that displays the time and I can ask her questions or play music.

For me, getting to sleep is less about blue light and more about apps designed to hold your attention and keep you there for as long as possible, through notifications and tantalising information. If sleep is precious, remove that distraction. Make it less tempting to while away half an hour when you should be going to sleep or chatting to your partner or reading a book. And if you reach for your phone first thing in the morning, your notifications dictate how you're going to start your day. It's good to take a moment to breathe and prep for the day, then start it the way you want to.

I track my sleep with a Fitbit Charge. It's useful for waking at the optimal time in your sleep cycle. I often have to wake early when I'm filming. It uses a vibration that won't wake your partner, which is thoughtful to your plus one. Also, I think Calm is a brilliant app. It's got some really good sleep and guided meditations that help with relaxation and anxiety, getting you into the right head space. "

Well, well, well

A mixed kit bag of tech designed to raise your game, support your training, boost your grooming and care for your health and wellness

PELOTON BIKE+

Peloton is the best high-end home exercise bike as it comes with online classes. This model can adjust resistance automatically to follow the instructor and has Apple GymKit integration.

PHILIPS SONICARE 9900 PRESTIGE

For superior plaque removal and healthier gums, this top-notch sonic toothbrush adapts to your brushing technique and gives personalised feedback via the app.

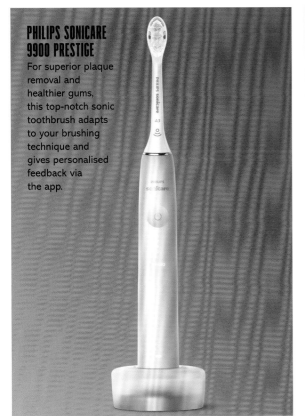

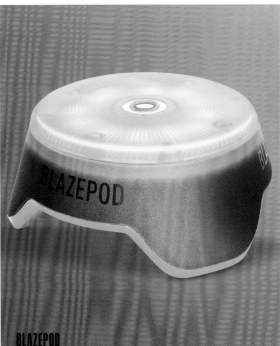

BLAZEPOD

Tap the wireless pods as they light up. Use them for exercise sessions in any sport, combining the fun of Whac-a-Mole with precise training stats. Coaches can pick drills from the app or create their own.

PETZL IKO CORE

This comfortable LED head torch weighs only 79g but puts out 500 lumens and has a rechargeable battery at the rear. It comes with a white storage pouch that can transform the torch into a lantern.

MORPHÉE

This French meditation gadget has no screen. Instead, you use the three golden keys to pick a theme, session and duration of meditation. There are 210 guided sessions designed to help you unwind and sleep.

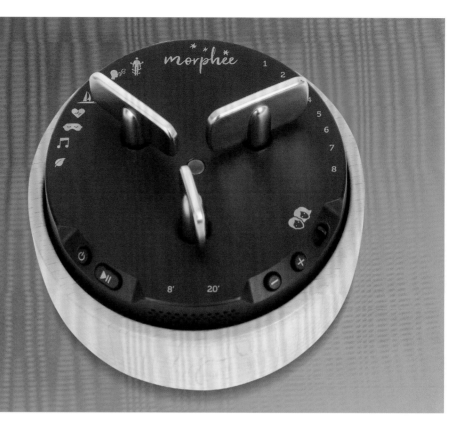

VAHA

Who needs Joe Wicks? This unique interactive sports mirror can display personal trainers and classes, as well as show your reflection. It's like having a virtual personal trainer in your home.

THE LITTLE BOTANICAL LIVING ART

A picture frame-style planter for biophilic art. Its reservoir waters the plants for two to three weeks and indicates when it needs refilling. It's made from recycled plastic and is supplied with houseplants.

BLUEAIR HEALTHPROTECT

This range of air purifiers comes in sizes to suit your space, cleaning the air of invisible particulates, dust and allergens, as well as bacteria and viruses. Germs are captured and killed, not recirculated. It's quiet and can be controlled by app or voice (Alexa or Google Assistant).

SCUBAPRO GALILEO HUD

A heads-up display (HUD) that attaches to the outside of your Scubapro mask. With minimal disruption to vision, the micro OLED display puts all essential data in front of you, making your dive safer.

FLITEBOARD SERIES 2

Imagine a surfboard that flies above the water. This electric board uses a hydrofoil below the surface to lift it, cutting drag and turning heads. The ride is thrilling, steering is intuitive, effort is minimal.

ORTIS DELEY

THE ACTION MAN

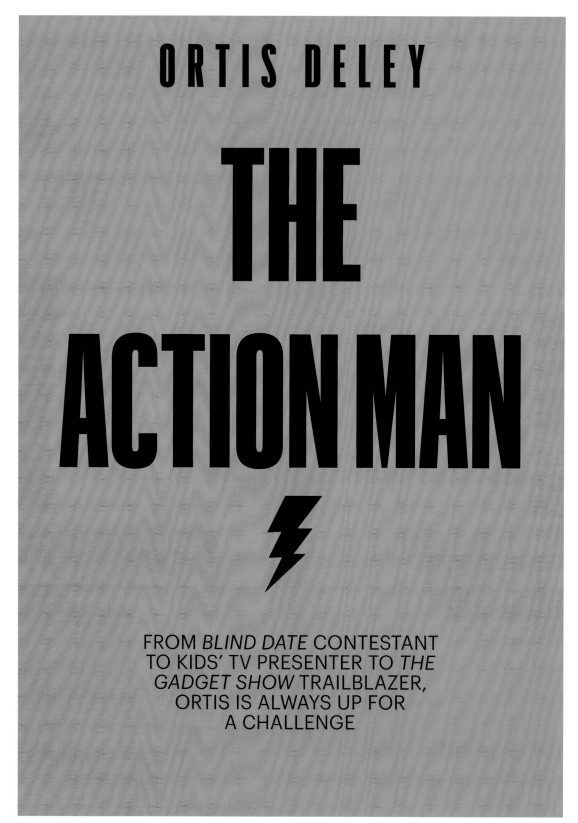

FROM *BLIND DATE* CONTESTANT TO KIDS' TV PRESENTER TO *THE GADGET SHOW* TRAILBLAZER, ORTIS IS ALWAYS UP FOR A CHALLENGE

South Londoner Ortis Deley made his TV debut in 1994 as a contestant on *Blind Date*, while he was studying pharmacy at the University of Sunderland. It led to a career as a TV presenter, starting on LWT (London Weekend Television), before moving to the children's TV channel Trouble and then a lengthy spell presenting programmes for BBC and CBBC. He presented Channel 5's *Police Interceptors* and Channel 4's sports coverage.

Ortis also has a number of notable credits as an actor, with roles in *Kidulthood*, *The Bill*, *Doctors* and – most curiously – in the 2005 Hollywood movie *Derailed*, alongside Jennifer Aniston, Clive Owen, Giancarlo Esposito, RZA and Xzibit. He started presenting *The Gadget Show* in 2009 and has been a regular since 2014.

Ortis also co-hosts a podcast called *We Build Too* – on black technology luminaries – and an online video show on Instagram TV entitled *Glassmates*, in which he and two good friends indulge their love of fine drinks. He lives in Surrey with his wife Rachel and their two young children.

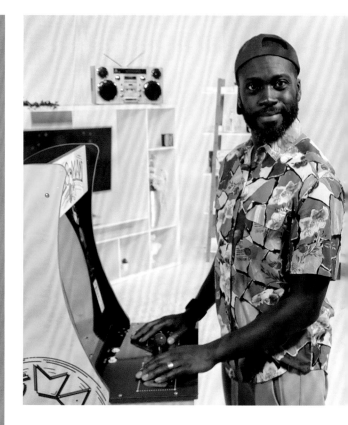

"I STUDIED SCIENCES ... AND I JOINED THE SHOW JUST AS CONSUMER SCIENCE AND TECHNOLOGY WAS GETTING REALLY EXCITING"

How were you a fit for *The Gadget Show*?
I studied sciences and I've always followed developments in technology, so *The Gadget Show* was the perfect fit. I joined the show just as consumer science and technology was getting really exciting. Every girl and guy wanted the latest phone, they wanted a laptop that worked properly, they wanted to increase the speed of their Wi-Fi. When I started on *The Gadget Show*, your typical bloke might have researched his TV, his phone, his computer and maybe a separate camera. Over the years, that has expanded massively – people are now researching all kinds of technology. And there are probably more women than men who are into tech now. Tech is taking over the house, with smart speakers and suchlike. If you look at the Internet of Things, you've got fridge freezers that can talk to cookers and lights that can talk to heating systems and so on. It's grown exponentially.

A lot of the studio sections seem to feature you being bullied by Craig...
Yes, we all decided that would be nice for Craig. It helps him feel important and it's also very beneficial for me to help the aged in any way I can. So, if I have to take a few knocks from Craig to make an elderly man

Above and left:
Ortis in 1994 on
Blind Date, and
as a presenter on
The Gadget Show

"AS SOMEONE WHO WAS A MASSIVE *RED DWARF* FAN IN MY TEENS AND TWENTIES, IT WAS A BIG THRILL WHEN CRAIG JOINED"

ORTIS DELEY

Debut on *The Gadget Show*:
30 January 2009

Home is: **Redhill, Surrey**

Go-to transport: **Bike**

Did you know? **Ortis made his TV debut on *Blind Date* in 1994**

feel good, I'll happily do that. Ha ha! I'm sure he'll flip when he reads that – or rather when someone reads it to him!

No, it's actually a friendly environment and we all get along really well. We've got to the point where less and less of the studio scenes are scripted – we can just riff off each other. Less of it is left on the cutting-room floor. The chemistry seems to work. And, as someone who was a massive *Red Dwarf* fan in my teens and twenties, it was a big thrill when Craig joined the show.

What was the most useless gadget you've tried out?
I remember trying out a flip-flop that doubled up as a water flask. There was a valve on the side of the sole that you filled with water. So, you could walk into the desert and have a few millilitres of water to drink after several hours. Really? It was an insane idea!

What do you love about gadgets?
I absolutely love observing how technology evolves so quickly. You go to CES – the annual Consumer Electronics Show in Las Vegas – and you see big firms like Google taking over half of Vegas to showcase what they've got going on. You also meet hundreds of small firms – men and women who've invented their own electric trikes and water propulsion systems and 'emotion-recognition robots' and whatnot. These guys aren't just thinking outside the box, they're operating on a different plane.

Having studied pharmacy, are you particularly drawn to health and medical technology?
Oh, most definitely. For instance, we discovered a French company that was working on dyslexia. Their research suggested that dyslexics have two dominant eyes rather than the norm which is to have one dominant eye, which is why they have difficulty reading. The two eyes are competing for dominance, so the brain can't decipher the symbols – words tend to jump around and shimmer. So, this

company had developed a light which flickered. You could adjust the flicker to help 'steady' those words and keep them still. And we discovered that dyslexics really could read with the assistance of this lamp. It's quite remarkable.

Do you really hold a Guinness World Record in cycling?
Yes, I've got a plaque on my wall from Guinness World Records to prove it! That was another scary experience. It was an ordinary road bike with dropped handlebars and they strapped six electric ducted fans onto it. They served as mini jet engines, powered by a battery. What they do is they give you a lot of wind assistance. The Guinness World Records guys turned up and said they'd give us a world record for the fastest wind-assisted cycle ride if we could break 62mph. So, we went to a drag strip where I got to 30mph and then started wobbling. We realised that the drag strip was sticky and this was affecting the bike, so I said the only way we're going to get the record is if we use the run off, which isn't sticky; it's where the cars slow down. It was filled with debris, so we needed to clear it. Then my speeds went up – 47mph, 56mph, and the producer then said to give it one more shot. I made it to 72mph but you can hear the dialogue – I was absolutely bricking myself! Anyone who has ever ridden a bicycle will know that 20mph is quick, 30 is fast. Anything over 50mph is absolutely terrifying!

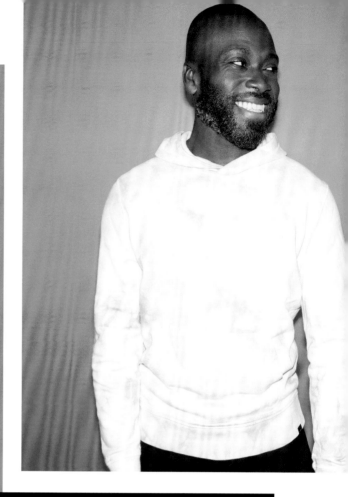

"FRIENDS AND COLLEAGUES FOUND IT HILARIOUS WHEN I SWITCHED, QUITE RECENTLY, FROM ANDROID TO APPLE. IT WAS LIKE I'D SWITCHED FROM CELTIC TO RANGERS OR SOMETHING!"

Are you a big cycling fan?
I do love cycling. It's my favourite form of transport. I live on the doorstep of the Surrey Hills, so it's easy to get a nice little workout. I could only afford to go bottom-end carbon, so I bought myself a Specialized Tarmac SL2 with a carbon frame a few years ago. I've since upgraded the wheels for something really good and lightweight. You start to realise that there's a whole wealth of tech and add-ons for bikes. You start looking for shoes and cleats and you have no idea where to start!

You have young children – how do you control their use of technology?
It's a tricky one. Despite being immersed in tech, I try to limit it with my kids. They get a small amount of screen time, but most of the time they play with old-fashioned toys to develop motor skills and so on. I also encourage them to play in the garden, in all weathers: they have a mud kitchen, a

Above and left: Ortis with Jon, and in action on his record-breaking, 72mph wind-assisted bike ride

trampoline and lots of building toys. And my wife and I are big on not using electronic devices at the dinner table.

What tech can't you live without?
It's probably my tablet. I have an Apple iPad Air 2. It's about five years old, but I use it all the time. I use it as my script reader, my teleprompter, my work diary. I review my photos on there, I read news on it, I read comics on it, I take it to the bath and the toilet! And I have an Apple Watch, which I wear most of the time. It's great because it enables me to use a certain fitness app but, in honesty, the battery is terrible! It's nowhere near as good as other wearables I've used, such as a Fitbit Charge 4 or a Garmin. Friends and colleagues also found it hilarious when I switched, quite recently, from Android to Apple. It was like I'd switched from Celtic to Rangers or something!

ORTIS TAKES FLIGHT

Flying has always been humanity's biggest ambition. On *The Gadget Show*, we've tried hoverboards that consisted of wooden boards linked to leaf blowers and whatnot. But it's such a difficult form of energy to control and steer.

The closest we came to a functioning hoverboard was when I went to San Francisco. A company had developed a magnetic floor and the board had electromagnets embedded into the base. The board was super thick; it looked like a comedy skateboard, but it was amazing to use – effortless, frictionless and so, so smooth.

I also learned to fly a plane on *The Gadget Show*, using flight simulators. Then I was put in a two-seater [pictured] and had to taxi, take off, do a little circuit and then land. It was exhilarating but very scary. Coming in to land is terrifying. If you come in at the wrong angle you could shear the plane in half.

If you see the clip, you'll see the colour return to the face of my co-pilot, who sat on her hands while I did it. When I landed, I filled the cockpit with expletives. I was jubilant. I'd touched down and nobody had gotten hurt!
See page 45 for more on Ortis in flight

Commander Spock adds an artefact to the ship's archive in 2016's *Star Trek Beyond*

SCIENCE FICTION

MOVIES AND TV SHOWS HAVE PREDICTED, EVEN INSPIRED, THE TECHNOLOGY OF TODAY AND TOMORROW. WE LOOK AT HOW EVERYTHING FROM *THE JETSONS* TO *MINORITY REPORT* GOT IT RIGHT

SCIENCE FACT!

STAR TREK'S TRICORDER

The original Gene Roddenberry TV series, which ran from 1966, is set in the 23rd century. The impact of its tech predictions has been widespread. We don't yet have teleports, warp-speed travel or even automatic sliding doors that go 'shhhhk', but we do have technology much like the tricorder.

The tricorder was a portable gadget worn from a shoulder strap. Its functions were to scan, record and analyse data. There were three models: standard, medical and engineering.

All three have inspired today's portable tech, but the medical tricorder has been of particular inspiration. In 2011, the Qualcomm Tricorder XPRIZE asked entrants to build a portable, non-invasive health diagnostics system that could autonomously diagnose a dozen common conditions.

No entrant completely fulfilled the brief, but millions of dollars were awarded to the best three. Basil Leaf Technologies came top with DxtER, a 3D-printed ball that packed in sensors and algorithms for diagnosing 34 conditions. It's based on analysis of patient data and years of experience in clinical emergency medicine and is still under development.

Meanwhile, cutting-edge kits turn your smartphone into a tricorder. Healthy.io makes Minuteful home test kits for kidney function, UTIs and antenatal monitoring, as well as a kit to help healthcare professionals track wound healing. Test results are fast and shared with clinicians.

Right: A *Star Trek* portable tricorder

Below: Healthy.io's Minuteful Kidney home test kit

Star Trek's Lieutenant Uhura using an earpiece for communications on the starship USS *Enterprise*

STAR TREK'S EARPIECE

Lieutenant Uhura was communications officer on the USS *Enterprise* in the original *Star Trek* TV series. While working on the ship's deck, she sported a single silver earpiece.

The earpiece was futuristic at the time but Bluetooth wireless earbuds are now commonplace. High-end models like the Sony WF-1000XM4 fit a bigger and better sound into ever smaller packages.

The earbuds work with Sony's LDAC file format, which supports high-res audio streaming over Bluetooth. A snug fit and active noise cancellation block out ambient sound. Four microphones mean you can make and take calls and use voice assistants, too.

Uhura would find the Speak-to-Chat mode useful. Simply start talking and your music pauses automatically; the earbuds let in ambient sound. The music resumes when your chat is over. Meanwhile, their innovative Adaptive Sound Control can learn to recognise places that you frequently visit and tailor sound to suit the situation.

Above: Sony's WF-1000XM4 wireless earbuds are smart listeners, with a Speak-to-Chat mode that reacts to your voice

BACK TO THE FUTURE'S HOVERBOARD

Back to the Future Part II came out in 1989 but was set in 1985 and 2015. It got technological predictions right in many ways. Nike has since produced self-lacing trainers; flat TVs are everywhere; and during lockdown there was even news footage of a drone walking a dog.

One of the film's most memorable predictions was the hoverboard. Gadget lovers are forever frustrated by the fact that we still don't have the hoverboards we were promised. The fact that Segway-style, self-balancing, wheeled scooters are often called hoverboards just makes it worse.

There have been many attempts at making a working hoverboard but the best have used magnetic levitation. Car maker Lexus created the Slide, which runs on a magnetic track. The 11.5kg board uses liquid nitrogen to cool superconducting blocks to -197°C, at which temperature they produce a maglev-like opposing magnetic field that makes the board hover above the track.

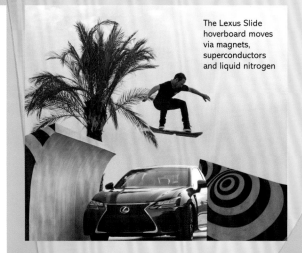

The Lexus Slide hoverboard moves via magnets, superconductors and liquid nitrogen

BACK TO THE FUTURE PART II GOT TECHNOLOGICAL PREDICTIONS RIGHT IN MANY WAYS. NIKE HAS SINCE PRODUCED SELF-LACING TRAINERS...

Marty McFly mastered the hoverboard in *Back to the Future Part II*

HITCHHIKER'S GUIDE'S BABEL FISH

Douglas Adams's *Hitchhiker's Guide to the Galaxy* started as a 1978 BBC radio comedy, then became a book, TV series and film. The eponymous Guide is an all-knowing, talking electronic book like today's Apple and Android tablets. One of its absurdly brilliant ideas was the Babel fish, "probably the oddest thing in the universe". Put one in your ear canal and it telepathically translates any language.

Today's fish-free version, Timekettle WT2 Edge earbuds, incorporate artificial intelligence (AI) and knowledge of 40 languages and 93 dialects for instant voice translation.

Right: Timekettle WT2 Edge earbuds break down language barriers with two-way simultaneous translation

THE JETSONS' 3D-PRINTED FOOD

Hanna-Barbera's space-age, 1960s animated sitcom was set in 2062. It predicted the flying car (we're still waiting) and a holographic Christmas tree. We don't have domestic holograms yet, but we did have US rapper Tupac's eerie holographic performance from beyond the grave in 2012.

The Jetsons also had a fabulous machine that could 3D print food. Press a button and it would fly onto a plate. 3D food printing is possible now. Hershey's has experimented with 3D printing chocolate and Natural Machines makes a printer called… Foodini.

Right: 3D food printers can make intricate custom-made shapes

CRAIG TALKS RED DWARF TECH

The Gadget Show's Craig Charles on how a fantasy future has become present-day reality

'There were throwaway jokes on *Red Dwarf* that are now no longer science fiction but science fact. Smart watches, virtual reality (VR) goggles, sat-nav, robot suitcases that walk alongside you, robot vacuum cleaners that you can order to clean up after you, genetically modified DNA – all these things are actually real.

In the episode 'M-Corp' we get a software upgrade and we can only see products made by the new company. So, there's a fridge full of Leopard lager but I can't see it. I can only see the M-Corp lager because of a perception filter. And I can't see my bed, so it looks like I'm lying in mid-air. I often think of that nowadays. We increasingly don't own applications – we rent them and they can disappear at any time!'

MINORITY REPORT'S G-SPEAK

The 2002 pre-crime thriller *Minority Report* was set in 2054 and is best remembered for its compelling user interface technology. Tom Cruise and colleagues wore special gloves to manipulate huge displays. They used intuitive gestures to move images, zoom in and out, and throw them onto other surfaces, including walls and tables.

The film's science adviser, John Underkoffler, was an MIT PhD research student when he took the call to go to Hollywood. Years later, inspired by the popularity of the on-screen tech, he went on to create in real life the technology that he had dreamed up for the film. Underkoffler co-founded Oblong Industries and developed g-speak gesture control. The company now champions its video-conferencing system Mezzanine, which uses g-speak but replaces the gloves with a sophisticated wand that can control the displays in three dimensions.

Right: Oblong Technologies was founded by John Underkoffler, *Minority Report*'s science adviser

This photo and right: HAL and today's less sinister AI, the voice-controlled Amazon Echo Dot

2001: A SPACE ODYSSEY'S HAL

There are no prizes for guessing the year in which most of *2001: A Space Odyssey* was set. Stanley Kubrick's elegant vision of future technology, including space travel, has shaped science fiction and science fact ever since its 1968 release.

Dr David Bowman is the human lead on a flight to Jupiter. HAL 9000, a talking artificially intelligent (AI) computer, is cast opposite him.

Ultimately, HAL does what we all fear AIs will do: he decides he knows better than the humans and starts taking over.

Our own AIs don't turn off human life support just yet, but anyone who has shouted repeatedly at Alexa (or Siri or Google) when it's wilfully misunderstood a very clear command will sympathise with Dave and shudder at the prospect of them replying: "I'm afraid I can't do that."

THE HI-TECH HOME OF THE NEAR FUTURE WILL MAKE LIFE EASIER, HEALTHIER AND HAPPIER. EXPECT EVERYTHING FROM ROBOTIC CHEFS TO SMART FURNITURE THAT REARRANGES ITSELF

THE NEW

NORMAL

he cult 1989 movie sequel *Back to the Future Part II* sees the director Robert Zemeckis having a stab at imagining the average home in 2015. Yes, the fax machine was a mistake, but he did get some things spot on, like huge wall-hung TVs, video calling, voice-controlled appliances and a house that knows when you're home. But our home of the near future is far more sci-fi than Zemeckis could have imagined, with TVs that unroll seamlessly from worktops, robot arms that'll dice and slice like a Michelin-starred chef, and toilets that can even detect illness before your doctor. Fact has finally overtaken fiction and we can't wait.

With their intuitive and attractively designed smart home gadgets, brands such as Nest, Hue and Ring have shot from niche start-ups to tech powerhouses, with consumers keen to enjoy the benefits of app-controlled heating, lighting and home security. And with 200 million voice-control speakers already sold, and an estimated 179 million homes in Europe and North America turning 'smart' by 2024, it's no longer just the early adopters. The smart home has gone mainstream.

But what really makes a home smart in the roaring 2020s? What does it add to the way you live your life, and what can we really expect in the near future?

Unglamorous as it is, the most important part of any smart home is fast, reliable Wi-Fi and, until 5G (or 6G or Li-Fi) helps to ease our router pain, ensuring our homes can handle dozens of data-hungry devices remains paramount. But for argument's sake, and so we can fast track to the fun stuff, let's assume everyone is blessed with bombproof broadband.

App-controlled lighting remains top of most smart home wish lists. Asking Alexa to dim the lights has obvious appeal, as does integration with security systems, the ability to use them to wake up gently, or simply create a different ambiance from millions of colours. The beauty of systems such as Hue and LIFX is that they're

starting to think for themselves – adjusting brightness depending on the time of day, for instance – and this will only improve in the next few years, with our homes learning and adapting to our daily routines.

Home security used to mean either an expensive, professionally installed alarm system or a big dog, but smart products from brands such as Arlo, Canary, SimpliSafe, Yale, Nest, Ring, Eufy, Hive and Ezviz have revolutionised how we monitor our houses with sensors, cameras, doorbells and alarms. For just a few hundred pounds you can have every angle covered, install facial recognition cameras, window sensors, robot vacuums with hidden cameras, and smart doorbells to enjoy dystopian levels of surveillance.

SMART FRIDGES CAN ADD INGREDIENTS TO YOUR ONLINE SHOPPING ORDER AND EVEN SUGGEST RECIPES TO AVOID FOOD WASTE

That said, it can be a confusing mishmash of kit, but a joint venture should help unify the smart home industry. More than 170 companies are involved, including Apple, Samsung, Amazon and Google. "The Matter alliance is helping to create connections between a wider range of products and brands," explains Kevin Spencer,

Smart-home security system SimpliSafe

Head of Product Management at Yale UK. "It will create an enhanced, connected experience." So, in theory, an Amazon Echo Show display could work seamlessly with a Google Nest Doorbell. It's what consumers want: to be able to mix and match.

With the Nest Learning Thermostat, home heating almost became sexy. Almost. Smart heating offers unrivalled app control and supreme energy efficiency. If you install smart radiator valves (try Honeywell Evohome, Hive or Wiser) you can control the heat in individual rooms and micromanage your home's heating and cooling. But change is coming fast, especially as gas boilers are being phased out for new homes by 2025. Expect a combination of smart electric radiators, air-source heat pumps, solar panels, home batteries and smart EV (all-electric vehicles) charging for your electric car and ebike.

Meanwhile, home appliances will get ever smarter. Early connected appliances simply offered app control, but as artificial intelligence (AI) has improved, they're finally becoming useful. Fridges like the Samsung Family Hub have internal webcams making shopping lists (almost) redundant, while some can alert you to food that's about to expire, add ingredients to your online shopping order and suggest recipes to avoid food waste.

Washing machines can self-dose and even order detergent for you and suggest cycles depending on the dirtiness of the load. The latest

Previous pages: The Moley automated kitchen unit with robotic arms

Above: Charge Amps' electric vehicle charging equipment

Below: SwitchBot Bot adds smart control to vintage technology

WALLOP'S TOP TIP

Consumer advice from *The Gadget Show*'s Harry Wallop

GENERATION RENT

"With apps like Fat Llama, you can make money by renting out gadgets you aren't using. Or save money by renting stuff rather than buying it new. It's like Airbnb but for gadgets."

Home fitness classes with the Vaha smart mirror

hobs (try the Home Connect range from Bosch, Siemens and Neff) can chat wirelessly to extractor fans for optimal results, while ovens can be preheated as you head home from work. Naturally, all this can now be controlled via voice assistants, and while there still isn't a robot that collects the laundry and loads the washer for you, at least you can ask Alexa how long is left on the cycle without having to be in the kitchen.

Speaking of robots, thankfully they will play a big part in the smart home of the future. Smart vacuum cleaners like Roborock can mop as well as vacuum; robot lawn mowers are finally reaching affordable levels; and once bonkers tablet-on-wheels designs like Temi, the personal robot, are becoming genuinely useful helpers.

Ori Living, in collaboration with Ikea, has designed a prototype for robotic furniture that changes use depending on your needs, switching from living room to bedroom to home office, hiding sections away when not needed. Moley, the robotic kitchen, promises an end to the tedium of chopping onions as its two anthropomorphic robotic hands dice, stir and whisk.

ONCE BONKERS TABLET-ON-WHEELS DESIGNS LIKE TEMI, THE PERSONAL ROBOT, ARE BECOMING GENUINELY USEFUL HELPERS

But enough about infrastructure, what about entertainment? With most TVs now offering smart apps, screen quality and design will dominate. LG Signature's OLED R rollable TV unfurls elegantly from a stylish sideboard, remaining invisible until needed, while C SEED has a gargantuan folding 165-inch 4K screen that raises effortlessly from beneath the floor, and Sony's Crystal LED, a 790-inch display boasting 16K resolution, is one step closer to us having fully interactive walls. "Hey Google, paint the living room dusty pink!"

As for music, even a basic voice-controlled speaker gives instant access to a lifetime's worth of music, and setting up a multi-room system is now simple, intuitive and inexpensive. New brand

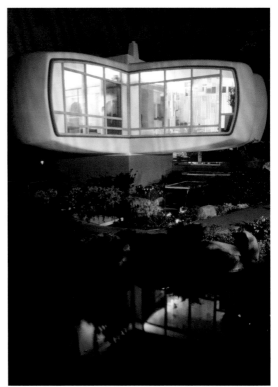

Below, left and right: Floating House, a CEDIA award-winning fully integrated home by Bridger Automation

Above left: Model of a home which featured in a US magazine in 1950 as a sustainable house of the future

Right: The Monsanto House, filled with gadgets, was an attraction at Disneyland in the 1960s

Zuma is making waves with its high-res streaming speaker and smart LED light combination that floods any room with light and audio. It's great for movies and parties, as well as meditation and wellness.

Health is almost certainly the next growth area for smart homes, with a slew of connected fitness gear, from exercise bikes like Peloton and interactive mirrors like Vaha offering strength, yoga and Pilates classes on-demand. But with a rapidly ageing population, fitness tech will become less about sweating and more about healthcare.

The Apple Watch is already able to track heart rate, check blood oxygen levels, monitor activity, sleep and notify emergency services if someone falls, but this health data will soon be available to share in real time with your doctor during a video consultation. Taking this a step further, Peter Aylett from the CEDIA Technology Council teases the gobsmackingly cool concept of micro-radar, which will see your smart home monitor you for good reasons. "A series of detectors and micro-radar will remotely monitor an individual's temperature, heart and respiration rate and detect biometric anomalies [read: heart attack]," he says.

WITH A RAPIDLY AGEING POPULATION, FITNESS TECH WILL BECOME LESS ABOUT SWEATING AND MORE ABOUT HEALTHCARE

This hyper-personal monitoring will also offer everyday smart home control. Peter adds: "People want control of conscious decisions such as what music to listen to, but will be happy to have subliminal decisions such as room temperature and light level made for them based on their own personal data."

Connected health tech won't just be for the boomers though. Expect smart toilets to become more popular in the coming years, especially as they'll soon be able to analyse your deposits for possible health problems. Meanwhile, the Poseidon smart mirror promises skin analysis and health tracking in the bathroom.

The smart home as we know it is still relatively dumb, despite the mountain of gadgets helping to make our lives easier, more efficient and cooler. But give it a few years and advances in AI will naturally anticipate your family's needs and the smart house will become a clever home.

GEORGIE AND ALEXA

The Gadget Show's GEORGIE BARRAT wants a heart-to-heart with her Alexa

The next phase of your Alexa, or other personal assistant, will arrange appointments for you and so much more. She has all your personal data, knows your likes and your dislikes and your search history and how you have handled situations previously. So, we will get to a point where we ask Alexa: should I take this job opportunity? Or should I dump that boy? Or what subject should I study at university?

I look forward to having in-depth heart-to-hearts with my Alexa. Computers like IBM Project Debater can already put together complex arguments, so I don't think it will be long before they'll be able to actually give us proper advice. Alexa would take your information and trends that are happening worldwide, explore the internet, profile-match you with people who have done similar things and give really good recommendations – for example, what holiday to go on.

CLEANER, GREENER...

HOW TO CLEAN THE AIR IN YOUR HOME IN RESPONSE TO EVERYTHING FROM COVID TO POLLUTION

Dirty air causes a staggering 36,000 premature deaths in the UK each year, according to Public Health England, with strong links to infant mortality, lung cancer and strokes. During the first Covid lockdown, outdoor air pollution rates were the lowest since records began, but indoors we have seen a rise in air pollution. Chemicals from building materials and furnishings, aerosol sprays and cleaning products all have the potential to cause harm.

Household dust has even been found to contain flame retardants, pesticides and particles from plastics and paints, but smart tech can help. Dyson's V15 Detect Absolute cordless vacuum is fitted with lasers to help illuminate invisible dust particles, while sensors measure dust down to microscopic levels, so you can see precisely how clean your floors are.

Air purifier technology has also evolved, with HEPA (high-efficiency particulate air) filters capturing over 99.6 per cent of particulate matter. Designs from Blueair feature air quality sensors that adjust power when levels rise and report air quality to an app. Similarly, the LightAir IonFlow range uses ionisation to destroy and prevent the spread of airborne viruses down to 0.1 micron; and the elegant Briiv harnesses the power of plants with its four renewable, biodegradable filters that it claims together have the same air-cleaning power as 3,043 houseplants.

But the single easiest way to improve indoor air quality is to open a window. Be strategic though: if you

Left, top: Velux Active integrates with windows and blinds for indoor climate control via a smartphone app

Left, bottom: Dyson's V15 vacuum cleaner can illuminate dust and automatically adapt suction power among other things

Below: Designed aesthetically for the home, the LightAir IonFlow air purifier can remove airborne and surface viruses

live near a busy road, keep the windows closed at peak traffic time, and minimise hay fever by keeping them shut in the morning when pollen counts are higher. Or Velux Active smart windows will do it for you, opening and closing on a schedule or according to temperature, sunlight, CO_2 levels and humidity for better air quality when it matters.

WFH in style

Does the new normal mean your dining table doubles as a desk? Here's how to work from home smarter, healthier and happier

1 GIGABYTE AERO 15 OLED
This powerful 15-inch laptop boasts the world's first X-Rite Pantone colour-calibrated OLED display, NVIDIA state-of-the-art graphics and a keyboard that illuminates in cute rainbow hues. The Windows laptop to buy if you're a creative.

2 TWELVE SOUTH CURVE
Your laptop is brilliant but it isn't designed for prolonged use at a desk or dining table. When WFH, use this lightweight metal stand to give your device an ergonomic lift and it won't be a literal pain in the neck.

3 NETGEAR ORBI WIFI 6
Fast, reliable broadband is everything. This mesh system offers max speeds, no Wi-Fi black spots and juggles demands so that Zoom doesn't drop out because someone else in the house is streaming a movie in 4K.

4 DYSON PURE COOL ME
Dyson's smallest fan and air purifier doesn't try to tackle the whole room. Instead, it creates a bubble of clean air perfect for your desk or bedside. It filters out allergens, pollution and airborne microbes.

5 JABRA EVOLVE2 85
This wireless headset, shown with charging stand, shuts out the world. It has active noise cancellation and 10 built-in microphones for outstanding call quality: it picks up your voice and no one else's.

6 LOGITECH ERGO K860
This supremely comfortable, wireless split keyboard lets you type with your arms in a more natural position. Pair it with your laptop for all-day working that doesn't cause wrist ache.

7 MOLESKINE SMART WRITING SET
Notebooks are great for brainwaves and shopping lists.

This one comes with special Anoto tech paper and a smart pen, so you can digitise notes and send them straight to an app for sharing or safekeeping. You'll never lose a note again.

8 VÉRITABLE CONNECT
This smart garden uses LED lights and silent hydroponic watering to grow your plants indoors three times faster than normal. It is infinitely easier than using a plant pot. The reservoir lasts three weeks and it alerts you via an app when it needs filling.

9 LOGITECH MX VERTICAL
Hold this advanced ergonomic mouse in a natural handshake position (right-handed only) and its unique 57-degree angle reduces pressure on your wrist. The button positioning is strangely intuitive, so it's easy to master.

10 PHILIPS 243B9H
A Full HD monitor with IPS panel for wide viewing angles. It has built-in speakers, a clever webcam that pops out of a slim bezel, and biometric sensors for Windows Hello facial recognition. A single USB-C cable connects to your laptop to charge it, minimising clutter.

11 PAPALOOK PA552
The ring light around this add-on webcam makes for a more flattering picture, whether that's for meetings or live streaming. Use the supplied tripod for a two-camera setup, or simply perch the webcam on top of your laptop or monitor.

12 JOY RESOLVE BARISIEUR
Caffeine boosts productivity and this does it in style, whether you use it to make fresh coffee at your desk or, thanks to the alarm clock, as a hipster bedside Teasmade. You can even add a wireless phone charger.

6

7

8

9

10

11

12

Seed capital

The future of food is to grow efficiently in cities with hi-tech, vertical farming

This is Pink Farms in São Paulo, Brazil. It takes its name from the artificial lights for its hydroponically grown vegetable plants. They're grown in space-saving vertical towers for 100 times more yield than farmland. That's right: 100 times more, not 100 per cent more. The process saves resources, too: it uses 95 per cent less water and 60 per cent less fertiliser than traditional growing methods. And the benefits don't stop there. São Paulo is the world's fourth most populous city. If you can grow food within cities, you can cut food miles dramatically and save waste. The greens leaving Pink Farms travel no more than 15km. They arrive fresh; nothing goes bad. Our cities of the future will include such locally and efficiently grown food, maybe hi-tech growing in our homes – turning food miles into food metres.

Get smart

Intelligent technology
for the smart home of
today and tomorrow

ARLO ESSENTIAL INDOOR CAMERA

This compact, wired camera
has a physical privacy cover
for when you don't want
family life to be watched.
Its affordable artificial
intelligence (AI)-based cloud
monitoring service alerts
you to unexpected activity
but ignores the cat.

MIELE G 7110 SC

This freestanding dishwasher measures dirt levels and automatically doses detergent, saving up to 30 per cent as well as cutting water use. It's Wi-Fi connected and can order new 1.4kg detergent refills when needed.

WUNDA WUNDASMART

A smart thermostat with a wireless E Ink control panel you can position anywhere. It works with Alexa and Google Home, while geolocation means the heating can turn on automatically when you're nearly home.

LENOVO SMART CLOCK 2

A stylish bedside gadget disguised as an alarm clock. Use Google Assistant or set reminders and alarms. The small colour touchscreen can display family photos, but to protect your privacy it comes without a camera.

IKEA TRÅDFRI

Use the Trådfri Gateway (right) to connect your Ikea motorised blinds, lighting or speakers, and programme this cheap-as-chips smart button (below) to trigger a series of actions. For example, turning off all the lights or getting set for movie night. It works with voice assistants, too.

NETGEAR MEURAL CANVAS II

Use this huge 19-inch-by-29-inch frame to display your favourite photos or digital versions of great art (read more about NFTs on page 183). Get Meural membership to choose from 30,000-plus pictures.

EKO AROMA

A simple, sleek touch-free soap dispenser. Just wave your hand under its motion sensor. You set the amount dispensed: 0.8ml, 1.4ml or 2ml. Rechargeable and refillable, it saves on plastic pump bottles.

PAPALOOK BM1

You can pan, tilt and zoom this high-definition baby monitor camera to keep a good eye on the sprog. Soothe them with the two-way talk and eight built-in lullabies. The cute bear design includes a memory card slot mouth and a thermometer tail.

NANOLEAF ESSENTIALS A60

Replace a standard light bulb with this app-controlled smart version and choose from more than 16 million colours. The bulb supports dimming and works with Bluetooth and Thread. It can be controlled by Apple HomeKit, Alexa and Google Assistant.

MORUS ZERO

This energy-saving mini clothes dryer is barely bigger than the drum because unlike normal dryers it uses a vacuum to make water evaporate at a lower temperature. It also dries gently but is four times faster and includes UV sterilisation.

Hansgrohe
RainTunes, a
multi-sensory
shower, with
video, light and
sound controlled
via an app

BATHROOM 2.0

THE BATHROOM OF THE
FUTURE IS SMART, COLOURFUL
AND WATER-SAVING. IT'S YOUR
PAMPERING PLACE BUT ALSO
YOUR PERSONAL ASSISTANT.
OH, AND IT LOOKS STUNNING

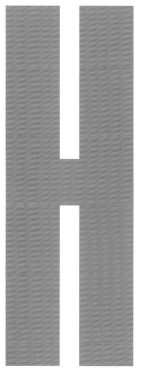

Have you ever experienced a Japanese toilet seat? If not, you should. They're packed with gadgetry, from personal entertainment to personal hygiene. And if the controls are in Japanese it's like playing Russian roulette: a button might play music, warm the seat, give you a surprise bidet or even a blow dry.

Toto is the market leader in Japan. Its latest loos, the Washlet RG and RG Lite, offer all of these luxuries but also more sophisticated self-cleaning functions, like the retracting wand jet that is cleansed with disinfecting electrolysed water, or a fine-mist spray that makes the bowl easier to scrub. Toto's top-of-the-range Neorest Actilight, which automatically opens as you approach, even has a UV light to kill bacteria. The remote control can save your favourite settings… and don't worry, it's in English.

Another 'shower toilet' worth a look is the Philippe Starck-designed SensoWash Starck for Duravit. Again, all the technology is hidden out of sight. Use the remote control or app to customise its hi-tech features, from the water spray intensity to the night light.

IMAGINE YOUR REFLECTION OVERLAID WITH SKINCARE RECOMMENDATIONS, THE NEWS HEADLINES AND YOUR PRIORITY EMAILS APPEARING AT A VOICE COMMAND

Left: Duravit's
SensoWash 'shower
toilet' has smart
functions, from seat
heating to descaling

Below: The
Simplehuman
Sensor Mirror
Hi-Fi not only plays
music but also
features voice and
motion control

Opposite: Kaldewei
creates the ultimate
bathing experience
with an audio
system concealed
in the bathtub

The humble bathroom mirror has the biggest potential
for a hi-tech makeover. Its entire surface could double as a
smart screen. Imagine your reflection overlaid with skincare
recommendations, the news headlines and your priority emails
appearing at a voice command, and the time and weather forecast
on hand in the bottom corner as you get ready for work.

VitrA has a concept smart mirror that charges your phone
wirelessly, streams music and uses Bluetooth to display smartphone
app notifications including weather and calendar. The
freestanding Simplehuman Sensor Mirror Hi-Fi is a smaller smart
mirror with a built-in speaker, Alexa voice control and motion
sensor so it lights up automatically as you get close. Its full-spectrum
ring light is designed for colour correcting make-up. Another
freestanding smart mirror, the HiMirror Mini 64G, on sale in the US
and Asia, also comes with Alexa voice control, as well as a touch
screen, skin analysis, advice and product recommendations. Imagine
an iPad and a vanity mirror have had a baby and you get the idea.

The smart bathroom of the future could monitor your health
unobtrusively, with smart scales tracking your weight and your
smart mirror noticing any physical changes before you do, alerting
you only when it matters. When you run low on meds or favourite
toiletries, the bathroom cabinet could simply order them itself.

Waterproof bathroom TVs never took off but listening to good
music in the bath is a must. The Kaldewei Sound Wave system uses

actuators concealed in any of its baths to turn the bath's flat surfaces into a giant speaker. The result is Bluetooth-streamed music that sounds stunning below, as well as above, the waterline.

Chromotherapy, or colour therapy, available in some high-end showerheads and taps, combines LED colour with light to stunning effect. For example, Crosswater's Revive Twist Wire Chromotherapy Shower Head has a remote control for you to select from a wide palette of colours to illuminate the 'rainfall', 'twist' or 'wire' water-flow modes.

Smartphone app or remote control are also ideal for getting your shower just so. It's an indictment of today's plumbing technology that when you first step in and turn on the shower it's probably freezing cold. British-made Triton's Host digital shower mixer changes that with a wireless controller, so you can start your shower and get the temperature perfect before you step in. You can even start it from outside a wet room. Presets allow for multiple favourite temperatures, or for different members of the household to have their own favourite.

If Japan is known for its gadgetry, Germany is the country to look to for high-end plumbing. The two big names, Hansgrohe and Grohe, sound similar for a reason. There's a rivalry between the company founded by Hans Grohe in 1901 and the one his son Friedrich set up when he left the family business in the 1930s.

Grohe's Rainshower 310 SmartConnect Cube comes with a waterproof wireless remote control, so you can get the shower started from outside but bring the controller in with you. Meanwhile, the premium Grohe AquaSymphony offers smartphone control of music, chromotherapy lighting and steam to accompany its six types of shower spray.

Hansgrohe's RainTunes is another stunning high-end digital shower system, combining water, light, sound and fragrance. Use a smartphone app, simple button or the RainPad control panel to select from preset moods or create your own.

Technology can save water and energy, and it puts you in control of it with precision. Grohe Plus taps, for example, have an integrated LED temperature display and hi-tech controls. You can run water at exactly the temperature you want, or wave your hand in front of the tap to switch between regular flow and an eco-friendly spray; it delivers a water-saving four litres a minute – perfect for handwashing.

And so we go full circle back to the loo, because technology can save water there, too. The VitrA Smart Panel toilet tells your app how much water it's using and can periodically flush itself to keep clean. It's not a waste of water if it saves on cleaning chemicals. You can control it via app or by voice command for touch-free flushing. Meanwhile, the VitrA V-Care range features remote controls that let you lift the seat without touching it.

All you need now is a designer shelf for your bog books... or better still, a special bathroom Kindle loaded with funnies.

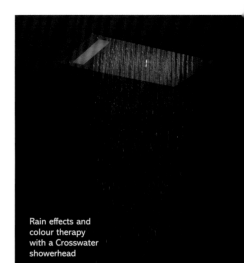

Rain effects and colour therapy with a Crosswater showerhead

THE PREMIUM GROHE AQUASYMPHONY OFFERS SMARTPHONE CONTROL OF MUSIC, CHROMOTHERAPY LIGHTING AND STEAM TO ACCOMPANY ITS SIX TYPES OF SPRAY

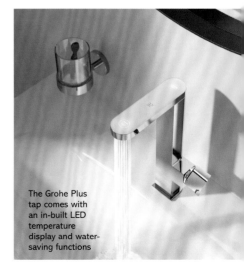

The Grohe Plus tap comes with an in-built LED temperature display and water-saving functions

ORTIS WANTS A BATHROOM THAT'S SMART

"Techie toilets might be a step too far. We Brits love toilet humour, but we won't discuss our… habits. Personally, I would love a toilet that talked to me, played music, massaged my butt cheeks, gave me a little wash and a blow dry. If I could afford those units, because they are hella expensive, I'd have one. I think they're fabulous.

Apps can already suggest make-up looks. That technology could be incorporated into your smart mirror. It could also give feedback on toothbrushing and analyse your skin. Rotating shelves could put your suggested medicines front and centre, then different ones when your partner uses the bathroom. A smart mirror could also update your medical records daily, even when there's no cause for concern.

And as a luxury, I'd have a bath with smart taps. You can ask your smart speaker to fill your bath: how deep and the exact temperature you like to soak in. That technology exists, I just can't justify the expense."

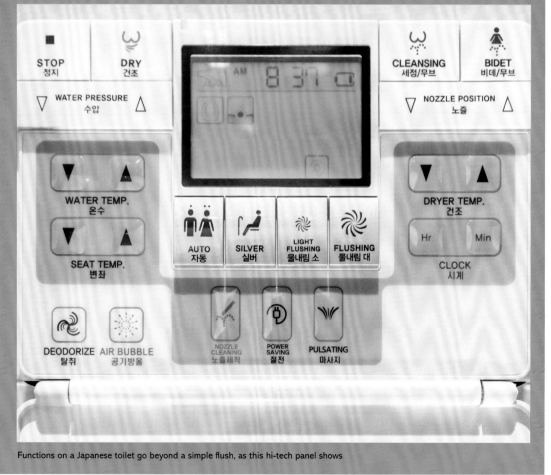

Functions on a Japanese toilet go beyond a simple flush, as this hi-tech panel shows

House specials

Eating in is the new eating out. These culinary gadgets will help you cook at home like a professional chef

OONI KARU 16

The Ooni pizza oven just got bigger and smarter. It burns wood or charcoal (or add a gas burner) and reaches 500°C in 15 minutes. A built-in digital thermometer tells you when it's ready to cook your huge 16-inch pizza.

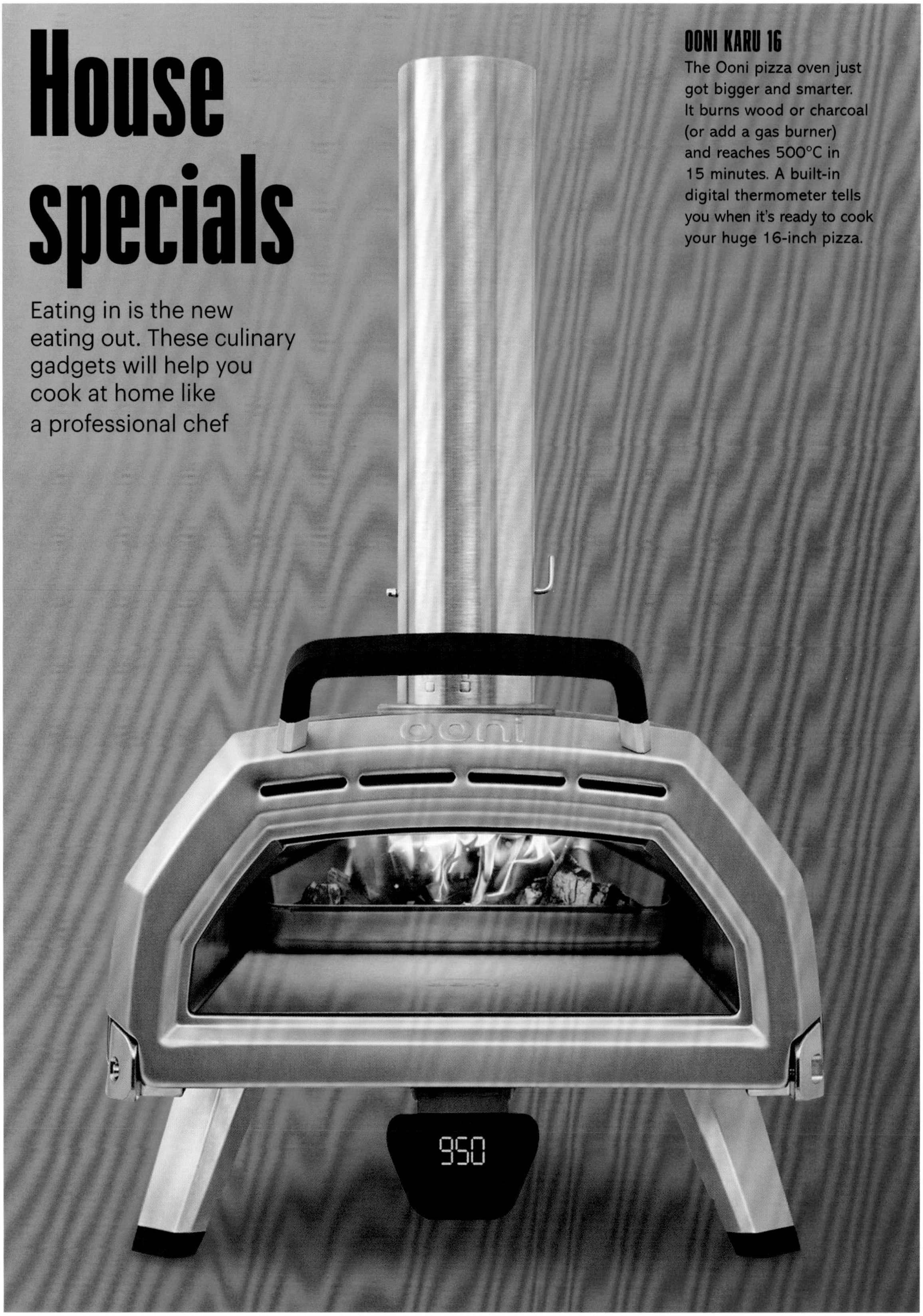

SAGE SMOKING GUN AND CLOCHE

A foodie gadget for adding smoky flavours at home.
It creates a cold smoke from small chips of wood,
including applewood and hickory, to infuse food
and drink. Make smoked cheese, fish, butter, garlic,
even popcorn and cocktails.

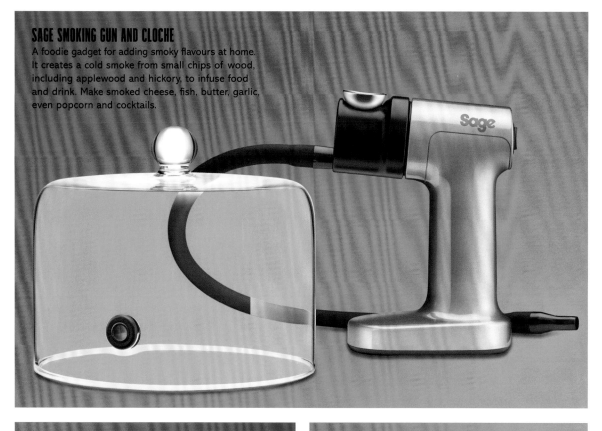

MEATER PLUS

Stick this pencil-sized
cordless probe in your meat
for the twin thermometers to
track the internal and oven
(or barbecue) temperatures.
A Bluetooth repeater in
the bamboo charger
gives a 50-metre
range. Use the app
to oversee cooking.

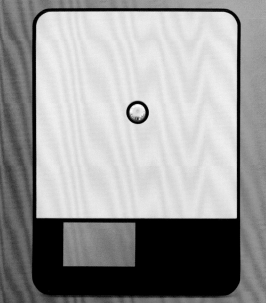

NJORI TEMPO

This smart induction single hob with built-in scales
comes with a temperature probe and water circulator
stored in its cork base. Together, they're perfect for
foodie cooking like sous vide.

SMARTER IKETTLE

This iKettle third generation comes with Alexa and Google voice control. Schedule it to boil for your breakfast brew or use geolocation (based on GPS phone location) so it boils just as you get home from work.

MINIBREW CRAFT

Brew your own beer from raw ingredients with this smart system. It mashes, boils and then cools the extracted wort, keeping it at the optimum fermentation temperature for a few weeks. Buy kits from micro breweries or use your own recipes.

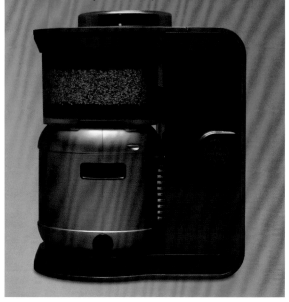

HESTON BLUMENTHAL PRECISION DUAL PLATFORM SCALES BY SALTER

Featuring a large platform that can weigh up to 10kg in 1g increments for big bakes, and a small one that can weigh up to 200g in 0.1g increments, for scientific precision. They measure liquid, too. All endorsed by the Michelin-starred Blumenthal.

CUISINART 3 IN 1 CORDLESS

This hand blender is USB rechargeable. Use it anywhere to blend, mash or whisk. No power cable means it's easier to clean in the sink and you'll never melt the cable on the side of a saucepan.

ISI GOURMET WHIP

Use nitrous oxide cannisters with the Gourmet Whip to conjure up sauces, soups, desserts and foams like a pro. The 25cl version is perfect for home use. iSi also makes the Nitro Whip dispenser, which uses nitrogen to make trendy nitro cold-brew coffee.

MAGIMIX COOK EXPERT

The classic French food processor does all the usual chopping and mixing but gets a 21st-century update. It now uses induction cooking to boil, steam, stir fry and bake in a big insulated stainless-steel bowl that also comes with it.

JON BENTLEY

THE BOFFIN

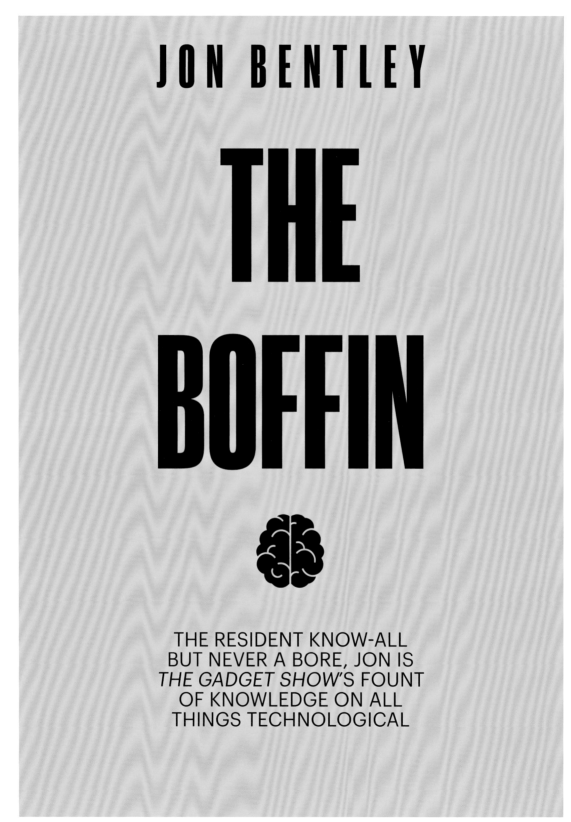

THE RESIDENT KNOW-ALL
BUT NEVER A BORE, JON IS
THE GADGET SHOW'S FOUNT
OF KNOWLEDGE ON ALL
THINGS TECHNOLOGICAL

The show's in-house professor, Jon Bentley was born in 1961 and studied geography at Oriel College, Oxford. His TV career started in the mid-1980s, as a researcher on *Top Gear*. He worked his way up to become editor, producer, series producer, executive producer and presenter. While *Top Gear* was being reinvented by the BBC in 2002, Jon followed most of the *Top Gear* production team onto Channel 5's *Fifth Gear*, and he joined the cast of *The Gadget Show* in 2004.

Jon also captained the Oriel College team on a celebrity edition of *University Challenge* in 2015, where he correctly identified an image of Gillian Anderson as being from a production of *Bleak House*. He lives in the south of Birmingham with his wife and two daughters.

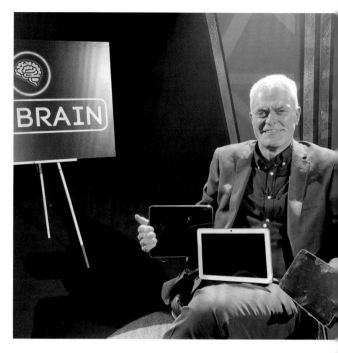

Above: Comparing tablets on *The Gadget Show*

Right: Jon (fourth from left) with the *Top Gear* crew in 1994

Have you always been a gadget fan?

As a child I was always obsessed with cars, but I was also interested in all sorts of household items. I built transistor radios from scratch – you used to be able to get these little box kits back in the 1970s. I used to be able to fix television sets, too. If it stopped working, you could open up the back of the set and usually you'd spot a broken valve or a leaking capacitor immediately, and I'd know usually know what to replace to repair it. Neighbours used to pay me to come over and repair their sets!

Do you still repair your household appliances today?

I do my best. I was particularly proud of successive repairs to our 24-year-old dishwasher. I soldered a poor connection, replaced the control board with another one and replaced a split tube. But leaks from under the machine defeated me. Instead of an expensive professional repair I replaced it with a new one. Probably the right decision.

Are new gadgets harder to repair?

You can't really mend microelectronics like a computer motherboard. The components are so impossibly tiny that you have to replace the whole thing. More annoyingly many modern gadgets, like my current laptop, are glued rather than screwed together. If it developed a faulty USB socket or defective battery I'd have to send it to recycling. That's not good enough. I welcome the increasing pressure on manufacturers to make products where larger components are replaceable so they can last longer.

You've written *Autopia: The Future of Cars*. What are your predictions for car technology?

Some enthusiasm is misplaced – self-driving expectations are being recalibrated, for example – and I don't think flying cars will ever become feasible. However, I can see eVTOL (electric vertical take-off and landing) vehicles, like giant drones, serving as premium taxi services. [Read more of Jon's car predictions on page 10.]

Are you sad that whole generations of classic cars may disappear from our roads due to emissions restrictions?
I don't think they will. Historic vehicles – including cars over 40 years old – are exempt from Clean Air Zone and Ultra Low Emissions Zone regulations in Britain. Classic cars represent such a small proportion of overall driving that they have little effect on total emissions. Like steam trains, the joy of seeing them outweighs any environmental damage. Some old cars are being converted to battery power while the rise of 3D printing is making old parts easier to make, too.

What tech evolutions have you seen since you started on *The Gadget Show*?
There's been so much, like video on demand, audio streaming, smartphones, Sky+ and HD TV, not to mention 4K and 8K. They've all emerged in the lifetime of *The Gadget Show*. Other technology has been launched, only to flop, and then surge back again. For instance, we covered newfangled video calling in the earliest programmes but it took the likes of Skype and FaceTime, and more internet bandwidth, to make the experience better before it took off. Other technologies have improved hugely but we're not quite there yet, like real-time translation. It feels close but we're still

"I USED TO FIX TV SETS. IF A TV STOPPED WORKING YOU COULD OPEN UP THE BACK AND SPOT A BROKEN VALVE OR LEAKING CAPACITOR"

some way from the fluency of the Babel Fish in Douglas Adams's *The Hitchhiker's Guide to the Galaxy*.

You're a bit of a camera buff, aren't you?
Oh yes. I loved doing film processing and printing back when I was at school and I still find taking pictures a joy. When *The Gadget Show* started, photography was in the throes of the digital revolution and cameras varied hugely in competence. More recently, cameras on phones have become better and better, and they can sometimes take better pictures than dedicated cameras. Their vast processing power means they're great at combining shots – for instance, they'll take multiple frames in quick succession in the dark and merge them to create one really good photograph. As with other areas of artificial intelligence, there are always going to be some problems: my iPhone sometimes assumes I want the sky to be blue, even when I'm shooting in fog and I want to record a moody and atmospheric misty effect.

JON BENTLEY

Debut on *The Gadget Show*:
7 June 2004

Home is: **Birmingham**

Go-to transport: **Car**

Did you know? **'Bentley Bend', the sixth corner of the *Top Gear* test track, is named after Jon**

You're also a music buff – what tech do you use?

I do listen to a lot of music: Radio 3, BBC 6 Music, classical, jazz, all sorts really. I love good wireless headphones: my Bowers & Wilkins P7s are a favourite, and I also love classic wired headphones like Sennheiser HD 25s and Beyerdynamic DT 150s. It's brilliant the way audio streaming has made so much music so accessible. And in-car entertainment has come on tremendously, largely because 4G and 5G internet signals are very good on most motorways. You really can listen to the world's radio, and any music you choose, in your car.

Have you ever got a world record?

We got a land speed record for driving a car that was powered by power tools. The frustrating thing was that the power tool motors didn't work together properly on the day. Although we did get the world record, it was only 75mph – we should, theoretically, have been able to do nearly twice that.

Will virtual reality (VR) take off?

You never forget your first virtual reality experience. For me it was a challenge where you had to cross between two tower blocks on a pole. Though I was fully aware my feet were on the floor of the NEC in Birmingham – on concrete – I still found the experience terrifying. VR hasn't really become mainstream. Wearing a headset is too much effort and it's disorientating. You're always accompanied by a perpetual slight dizziness. The rewards aren't really worth the inconvenience.

You've tested using tablets to make art. Have you done any yourself?

When the iPad Pro came out, I hoped it would kickstart my artistic career. I bought one with an Apple Pencil and started drawing and painting on it. But I haven't done enough. I'm still using the iPad six years later, which is a testament to the longevity of Apple's products, but mainly to read magazines and catch up with the news. I have managed to slash my consumption of physical newspapers and magazines.

Before *The Gadget Show*, you were a *Top Gear* producer. Can we blame Jeremy Clarkson on you?

Ha ha! I did give Jeremy his first TV job – I loved his very funny columns in *Performance Car* magazine and offered him a screen test for *Top Gear* in 1988. I can't take credit for the big reinvention of the show in 2003,

"IN-CAR ENTERTAINMENT HAS COME ON TREMENDOUSLY, LARGELY BECAUSE 4G AND 5G SIGNALS ARE VERY GOOD ON MOST MOTORWAYS"

Left: Kitted out for a world record attempt in a power tool-powered vehicle

Below: Jon where he loves to be – in the hot seat

but I like to think we laid the groundwork for it. That was very much the work of the producer Andy Wilman. I think he and Jeremy did a tremendous job.

How did you get into *Top Gear*?
I'd always been interested in cars so, after leaving university, I got a job with Ford. I was a bit bored one day, so I was looking through the Monday job advertisements in the *Guardian* and saw that *Top Gear* wanted researchers. I had enthusiasm and industry experience. I got the job and ended up becoming a researcher, then a director, then series producer, executive producer and presenter.

Do you prefer to be the presenter or the producer?
That's a difficult one but when you're presenting you have less to worry about. When you're a producer, you're under lots of stress because you feel responsible for everything. We're very lucky to have wonderful producers on *The Gadget Show*. Sometimes my inner producer will come out and I'll make suggestions, but you have to let them have the final say.

JON'S TOP FLOPS

Google Glass was always around the corner, but there was the social pariah status attached to it. You could be filming anyone you were looking at!

3D is an obvious thing that never really took off, although cinemas gave it a good try about a decade ago. I think it requires too much effort. The disorientation distracts us from the films and games, and the appeal wanes quite quickly. The thing I'm waiting for is achieving those fabled TV screens in contact lenses. Maybe then it will be like high-quality video, seamless and easy to incorporate into our world.

The worst things are forms of technology that simply don't work, or don't do what they're supposed to do. I tried out these wind turbines that you attached to the handlebars of your bicycle. They were supposed to charge the battery on your phone while you were cycling. I must have cycled 20-odd miles and it barely charged my phone at all!

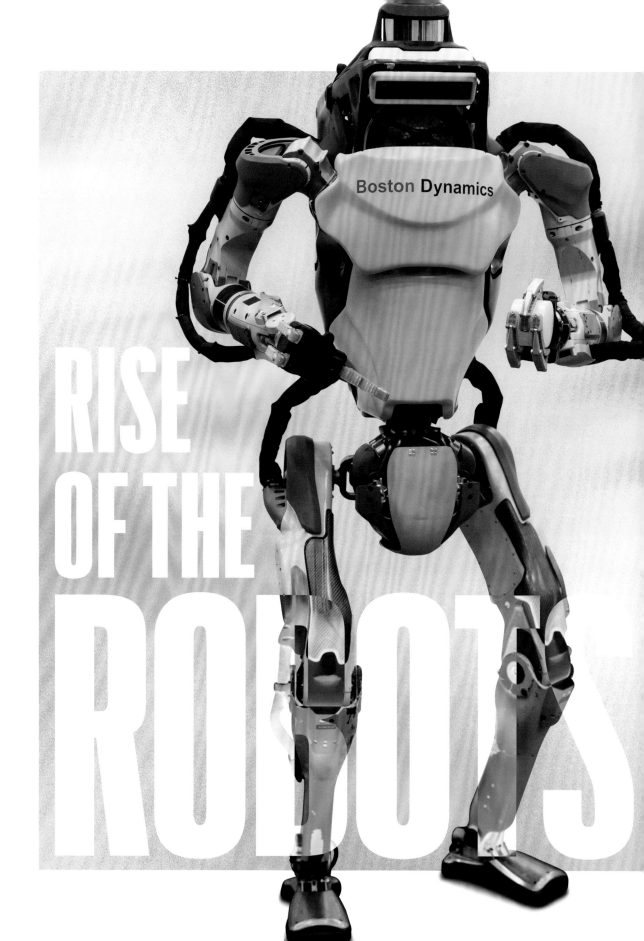

A NEW GENERATION OF ROBOTS WILL DO THE DOMESTIC DRUDGE AND THE MOST DANGEROUS JOBS. IS IT TIME TO WELCOME OUR NEW FRIENDS ELECTRIC?

Left: Valkyrie is a 6ft 2in-tall robonaut built by engineers at the Johnson Space Center for NASA

Opposite: Boston Dynamics' Atlas humanoid robot has 28 hydraulic joints, helping it to run and jump

oday vacuum cleaners, tomorrow the world! The development of new robots promises to revolutionise our homes and workplaces. But will they do the work while we put our feet up or will we work for them?

The most stunning robotic innovations come from the labs of Boston Dynamics, which was set up by an academic from MIT in the early 1990s. Its early robots were quadrupeds and made fascinating viewing, sometimes uncanny in their lifelike movements and sometimes skittering like new-born giraffes. There's a great video of the lightweight, dog-like 'Spot' robot slipping on banana peels. You can almost imagine it feeling embarrassed. But there's also video of it loading the dishwasher, giving a sense of how a sophisticated robot could one day be a pair of helping hands around the house.

The Boston Dynamics Atlas robot is even more uncanny because it's humanoid, a biped. It doesn't just walk; it dances and performs gymnastics. It looks like something from a science fiction movie but it's real. The obvious use for such sophisticated robots is military, but perhaps more for clearing minefields than for combat. Robots are ideal for environments that aren't human friendly. They're already used to inspect high-voltage equipment, defuse bombs, or even clean up radioactive dirt, such as inside the Fukushima reactors in Japan.

Not breathing, eating or sleeping makes robots ideal for space exploration, too. NASA has

Below left: MHI-MEISTeR robot, seen cutting a pipe, helps with nuclear accidents such as at Fukushima

Below: The inner workings of the Boston Dynamics Atlas robot

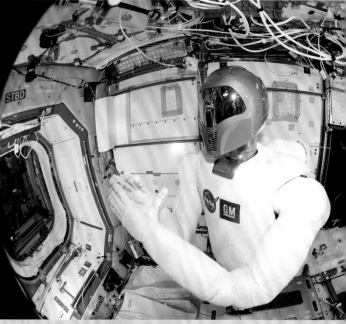

Above left: Ortis
with RoboThespian,
an entertainment
robot used in
settings from
theatres to TV

Above right:
NASA robonauts
have already
been sent to the
International
Space Station

developed the semi-autonomous robot Valkyrie, also known
as R5, to operate in hostile environments. The robonaut could
help build a base on the moon or Mars. Its human-like form
means it can use tools that were originally designed for
humans. And if an operator picks a destination, the robot
can plot its own safe path to it across difficult terrain.

Robots don't have to be humanoid though. NASA also has
a trio of cube-shaped Astrobee robots
in the International Space Station (ISS).
Called Honey, Queen and Bumble, they
use electric fans for propulsion and each
has an arm to grab handrails or hold
items. They help with jobs around the
ISS and return to a docking station
automatically to recharge their batteries.

Back on Earth, we can already buy
robots for the home. The most popular
domestic helper is the robotic vacuum
cleaner. The latest models are slim enough to clean under
most furniture and clever enough to sense a precipice, to stop
them falling down the stairs. Top of the iRobot range is the
Roomba s9+, with a 'base station' that automatically collects
the dirt when it docks. Instead of emptying it after each clean,
you can empty it monthly. You can also team the robot with
a companion, the Braava mop robot, to wash the floors.

Robotic lawnmowers are more expensive but potentially
more labour saving. While vacuuming is complicated (you
might have to move furniture and pick up toys to do it

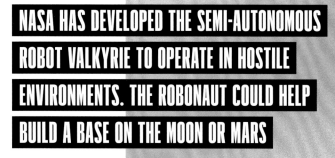

NASA HAS DEVELOPED THE SEMI-AUTONOMOUS ROBOT VALKYRIE TO OPERATE IN HOSTILE ENVIRONMENTS. THE ROBONAUT COULD HELP BUILD A BASE ON THE MOON OR MARS

properly), lawns are typically empty spaces ready to be mowed. The Honda Miimo HRM 70 Live can be scheduled to quietly mow during the week, controlled via an app or Alexa voice command, or it can intelligently schedule itself to mow based on weather forecasts and grass growth.

The Husqvarna Automower 405X and 415X are robotic mowers that also have voice control with Alexa or Google and are GPS assisted for improved navigation. Like the Honda though, you still need to install a thin perimeter wire to indicate where to mow. But Husqvarna's pro model, the Automower 550 EPOS, does away with this. It's big and expensive and more aimed at mowing football fields than gardens, but the tech points to a future where your robot will be intelligent enough to know where to mow and where not to.

But will robots do the weeding and pruning? Or do all of the cooking? Again, there's a glimpse of the near future in the Moley robot kitchen which features a pair of track-mounted robotic arms, teamed with appliances and cookware designed to be robo friendly and also human friendly. The Moley kitchens are aimed at high-end residential use but the UK start-up is looking into commercial kitchens, too.

Samsung also uses a robotic arm in its Bot Handy concept. It shuttles around the house on wheels and uses advanced artificial intelligence (AI) to recognise and pick up objects, judging how much force to use. You can even task it with chores like cleaning a messy room or loading the dishwasher.

For now though, home assistance robots are... armless. Temi looks more like a tablet on wheels and Amazon's secret home robot project, codenamed Vesta, is similar. It's reportedly like an Amazon Echo on wheels, complete with Alexa voice control.

Above: Honda's Miimo HRM 70 Live robotic lawnmower

Below, far left: Airwheel SR5, a self-driving suitcase equipped with anti-collision sensors

Below centre: The companion-like Samsung Bot Care5

Below: Samsung's single-armed Handy, which can pour a drink or load the dishwasher

One of the best practical uses of robotics is the Airwheel SR5 robot suitcase – clever carry-on luggage that can automatically follow you through the airport. It is wirelessly tethered to a smart band on your wrist and has sensors that prevent collisions.

Aside from following us through the airport, or doing the vacuuming and lawnmowing at home, robots will do some jobs currently performed by humans in employment. They'll typically be the monotonous jobs that we don't want, such as working on production lines and in warehouses, leaving humans to do the more interesting work. Industrialisation has always changed the job market, not become the boss of us.

It's naive to think that we will work less though. When women joined the workforce en masse, men didn't lose their jobs or go part time. We just all worked. Household incomes increased but so did expectations. We earned more and we spent more. And you'll want to earn more... not least so that you can afford to buy state-of-the-art domestic robot helpers.

THE AIRWHEEL SR5 ROBOT CARRY-ON SUITCASE CAN AUTOMATICALLY FOLLOW YOU THROUGH THE AIRPORT

CRAIG ON ROBOTS GOOD AND BAD

'Kryten from *Red Dwarf* is, of course, my favourite robot – he's the best in the world. He's like a cross between Herman Munster and C-3PO in many ways. He's a lovely robot. I think everyone needs a Kryten.

Everyone wants a domestic robot. To do your washing, mop your floors, stuff like that. They're good robots. And the dishwasher, I think that is a kind of robot. It just needs to develop a bit more so it can do your dishes and then put them away.

But I find some of the latest robots disturbing. The robots that Boston Dynamics is developing are completely frightening. They can do somersaults and land on their feet, run through walls. You can see that in a dystopian society we'd be having wars with robot soldiers. But to watch their dexterity, their hand movements, their precision... it's a wonder to behold. So, I find them frightening, fascinating and strangely kind of beautiful as well.'

Left: Kryten, left of Craig as Lister, in *Red Dwarf*

CONVERGENCE HAS TURNED THE PHONE IN YOUR POCKET INTO THE HUB OF EVERYTHING: THE KEY TO UNLOCK SMART HOME TECH, STREAMING, VOICE ASSISTANTS AND MORE. WHATEVER NEXT?

ONE THING TO RULE THEM ALL

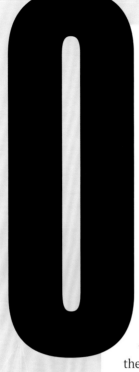

O n 9 January 2007, Apple founder and technology genius Steve Jobs went on stage to reveal a much-anticipated all-in-one gadget which he described as a "an iPod, a phone and an internet communicator".

Smartphones have ruled the world of technology ever since, now with more than a billion sales each year. Your phone can control everything from your television to your music system to your burglar alarm. But what's going to come next?

Some of the technologies which will augment (or replace) the phone in your pocket are already here. Others are still at the prototype stage but set to upgrade our lives in unimaginable ways.

PLUGGED IN

Brain-machine interfaces are perhaps the most dazzling (and terrifying) new way for us to interact with machines... using the power of our mind. In their most extreme form, brain-machine interfaces will allow us to connect directly to computer systems. Some believe they could even lead to immortality, with human beings becoming ghosts uploaded into online systems. In the nearer future, they could become a fast, reliable way to control computers – novel for everyone and offering new hope to paralysed people.

Tesla CEO Elon Musk claimed in 2021 that his brain-computer interface company Neuralink may be able to implant a 'neural lace' into a human brain later in the

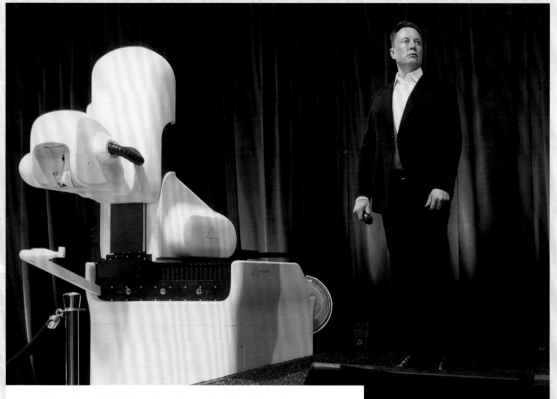

Above: Elon Musk
at a presentation
for his Neuralink
interface, which
aims to connect
brain and machine

Opposite: VR
integrated into
a bike helmet

year. The SpaceX billionaire showed off a video that
demonstrated how Neuralink technology allowed a monkey
to play video games using its mind. The start-up launched in
2016 and Musk has said that he hopes to test the technology
so that people with paralysis can regain independence.

Musk has spoken of a future where such devices will allow
people to 'merge' with machines. Less invasively, headphone-
style devices from companies like Neurable could 'read'
wearers' minds, switching off notifications from people's
smartphones when they're in a period of intense focus.

Other gadgets will add 'brain control' to just about anything
in the home, with the upcoming MyndHub letting you control
USB gadgets and smart home equipment with the power of
your mind – using an EEG (electroencephalograph) headband
to communicate mental focus and
calm to the machine to switch on
the lights (or even drive a remote-
control car).

RISE OF THE ROBOTS

Most of us are used to the idea of
robot helpers in the form of voice
assistants like Siri, but could real, physical robot servants
be about to arrive in our homes?

Amazon seems to be betting on the technology with
a home robot codenamed Vesta (named after the Roman

> **ELON MUSK HAS SPOKEN OF A FUTURE WHERE DEVICES WILL ALLOW PEOPLE TO 'MERGE' WITH MACHINES**

goddess of the hearth, not the matches), which reportedly resembles a trundling Echo Show on wheels.

Other robots are aiming for the same vibe. Temi, an Alexa-enabled tablet on wheels, can move under its own steam but be controlled from anywhere in the world. The makers pitch it as perfect for greeting guests at a hotel or looking after elderly people.

For those who have been dreaming of having a personal R2-D2 to actually do the chores, Samsung has a bold vision of the future with its Bot Handy, which scoots around the home on a vacuum cleaner-esque base and can pick up mess with its robotic arm. It can even serve you a glass of wine.

WEARABLES... AND HEARABLES

'Hearables' is a trend that's with us already thanks to Apple's all-conquering AirPods – headphones that offer a few smart functions.

But over the next few years, the 'hearables' in your ear could get a great deal smarter, adding functions such as heart rate monitors (it's easy to measure your heart rate from blood vessels in your ear), plus more voice-control functions, until you barely have to fish your phone out of your pocket.

Above: Personal robot Temi interacts with humans using AI technology and autonomous navigation

Cornell University in the US showed off a device which can 'read' people's facial expressions just from the contour of their cheeks, allowing wearers to transfer their smiles or scowls onto an emoji (or a video game character). The device, the ear-mounted C-Face, is sensitive enough that it even allows silent speech commands.

THE HEARABLES IN YOUR EAR COULD GET A GREAT DEAL SMARTER, ADDING HEART RATE MONITORS AND MORE VOICE-CONTROL FUNCTIONS

With Apple AirPods a runaway success, expect to see more rival devices such as Amazon's Echo Buds. These are more than just wireless headphones: there's a growing focus on talking to a built-in personal assistant. Walking down the road muttering to yourself will be very on trend for 2025. So, expect to see more devices that blur the lines between head and eye wear and entertainment devices, in a similar vein to Bose's Frames sunglasses (which have a Bluetooth music player discreetly built into their arm).

Wearables such as Apple Watch will just be the tip of the iceberg, as technology becomes integrated into everything from jewellery to the clothes we wear. The Oura smart ring already offers many of the functions of a smart watch – just on your finger – and gadgets such as Google's Jacquard offer electronics integrated into a Levi's jacket, allowing you to listen to music, send messages and take photos directly from a sleeve. Such 'smart clothing' is just another

WALLOP'S TOP TIP

Consumer advice from *The Gadget Show's* Harry Wallop

REFURBISHED PHONES

"Refurbished gadgets are far more environmentally friendly. Why fork out for a new piece of aluminium containing rare earth metals? But check the small print: are you getting an unlocked phone and a year's guarantee?"

way to let phones fade into the background – while staying connected to the digital world.

BEYOND REALITY

The next frontier for tech companies looks set to be augmented reality (AR) and virtual reality (VR). Apple is strongly rumoured to be working on a high-performance headset of its own, behind closed doors.

AR could soon add the sense of touch, with wearables such as the prototype TeslaSuit allowing people to 'feel' the virtual worlds they explore. This covers everything from bullet impacts to playing virtual games of dodgeball.

Facebook's Mark Zuckerberg has said that he hopes smart glasses will allow wearers to 'teleport' to other locations in VR, and that it could help climate change by reducing travel (but he admits that creating AR glasses is a huge technical challenge). Hardware such as Snapchat's Spectacles 2.0 is in the experimental stage but allows wearers to 'capture' the world in 3D.

Microsoft's HoloLens 2 mixed reality (MR) device has so far been aimed at businesses (and the US military). But the company is using it in a museum display, which brings extinct animals like sabre-toothed tigers to life, highlighting the potential power of AR and VR entertainment.

ALPHABET CITY

The Gadget Show's ORTIS DELEY explains VR, AR, MR and HUDs

" Virtual reality (VR) is full immersion. Augmented reality (AR) is when a layer of information is laid upon the real world. Mixed reality (MR) recognises that there are objects in the real world that you can interact with, as well as putting a layer of virtual interactivity over the top. For example, HoloLens. It brings what you're doing, where you are, to where I am, and vice versa. We can share a space together virtually but still see what's going on in each other's realm.

You've already got heads-up displays (HUDs) in some high-end cars. You can read speed gauges, petrol levels and maps, projected onto the windscreen. There are helmets that can do away with the dashboard in a fighter jet: all the information you need is on the heads-up display just in front of your eyes, but still allowing you to see what's going on in the outside world. I don't think it will be long before motorcyclists have them in their helmets. And we'll see more of it in cars, not just expensive cars but trickling down to more affordable ones.

Microsoft's HoloLens 2 is a mixed-reality device that enables the wearer to touch and move holograms

HOLOGRAMS AND TALKING TELLIES

Could the TV in your living room soon show off lifelike 3D images with no glasses – like Princess Leia in *Star Wars* saying, "Help me, Obi-Wan Kenobi, you're my only hope"?

Several companies, including Looking Glass Factory and IKIN, are pioneering glasses-free 3D technology which could deliver hologram-like images in people's living rooms and also allow people to capture and transmit themselves in 3D. This would make it easy to virtually visit friends or family anywhere in the world.

But that's not the only change coming to a living room near you. Samsung has already trialled 'portrait mode' televisions (vertical, to reflect the fact that most smartphone content doesn't come in landscape format) with its Sero TV, which can rotate to change format.

Smart displays – like smart speakers but with screens – are forecast to become ever more important in our homes, handling everything from recipe videos on YouTube to a growing army of gadgets such as smart doorbells. Over the next four years, the market for smart displays like the Google Hub is forecast to grow at a rate of 30 per cent year on year, according to Mordor Intelligence.

THE FUTURE OF PHONES

So what will the smartphone of the future look like? In stark contrast to the previous decade of smartphone evolution, it won't simply be just like your last one... but a little bigger.

GLASSES-FREE 3D TECHNOLOGY COULD DELIVER HOLOGRAM-LIKE IMAGES IN PEOPLE'S LIVING ROOMS

Above: Samsung's Sero smart TV has a rotating screen for vertical or horizontal viewing

Below: In the far future we could be operating devices at the touch of an arm

Folding phones are in their infancy now, but handsets such as the Galaxy Z Fold3 are increasingly well reviewed. Phones that unfold for bigger screen area (and a hard-wearing exterior) are set to be a big growth area. But will Apple introduce a folding handset? It certainly has patents relating to the idea – and analysts have suggested that it could launch as early as 2023. Where Apple leads, others will follow. Wait for a new buzzword, too, describing phones that fold out into tablets: the 'smartlet'.

The other improvements to the phone in your pocket will make it more and more like a featureless slab. Cameras will disappear under the screen (Chinese phone-maker Oppo has already showed off a handset that does this).

We will also see phones with no ports, at all, not even a charging port, as smartphone makers seek to streamline their devices still further. There's one already – the appropriately named Meizo Zero – which does everything wirelessly and the screen itself acts as a speaker and microphone. It's sleek, it's waterproof and the big brands should take note.

THE STORY SO FAR

FROM A SIMPLE ROLODEX TO A ONCE UBIQUITOUS BLACKBERRY – A POTTED HISTORY OF THE CONVERGING TECHNOLOGY THAT HAS LED TO THE MOTHER OF DEVICES, THE SMARTPHONE

ROLODEX 1958
The Rolodex quickly became ubiquitous in offices worldwide. It's a rotating wheel of cards on which users can write phone numbers for contacts (or just staple on a business card directly).

FILOFAX 1980
Filofax dates from 1921 but will forever be connected with the Yuppies. Revamped by Ian Logan in 1980, it was a paper organiser with removable sheets for events and contact details.

PSION ORGANISER 1984
Billed as the 'World's First Practical Pocket Computer', this had a six-by-six keyboard with the letters in alphabetical order rather than Qwerty, and could store contacts and do sums.

APPLE NEWTON 1993
For anyone who believes Apple can do no wrong, this device shows the company can make mistakes. The handheld Newton had handwriting recognition but was expensive and clunky.

IBM SIMON 1994
If you thought smartphones began in the 21st century, think again. This cumbersome IBM device combined basic computing functions with the ability to do calls and emails.

NOKIA COMMUNICATOR 1996
Nokia dabbled in 'personal organiser' territory with its vast Communicator devices, which came with Qwerty keyboards and the ability to send messages, emails and even faxes.

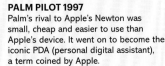

PALM PILOT 1997
Palm's rival to Apple's Newton was small, cheap and easier to use than Apple's device. It went on to become the iconic PDA (personal digital assistant), a term coined by Apple.

BLACKBERRY 2002
BlackBerry 5810, the company's first phone, focused on email and was a cult hit in the consumer and corporate worlds alike. The Qwerty keyboard resembled the blackberry fruit.

IPHONE 2007
The iPhone wasn't the first smartphone but it heralded the beginning of the end for the 'personal digital assistant'. The arrival of the App Store a year later hammered the final nail into the coffin.

Look who's talking

Make the most of
technology simply
by using your voice,
for intuitive hands-
free control

AMAZON ECHO DOT KIDS

The children's version
of Echo Dot 4th gen
– with Alexa voice
assistant built in –
looks cool, with a
tiger or panda
makeover. It comes
with a year of Amazon
Kids+ content and
has parental controls
enabled. Crucially, it
doesn't get annoyed
when children keep
asking "Why?"

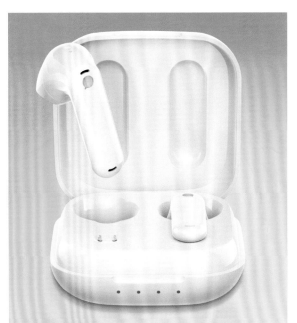

SONY SRS-NB10

An unusual wearable, worn like a collar and comfortable enough for all-day use. Speakers create a personal sound space and microphones let you use your phone's voice assistant. Work without shutting out the world.

OTTER.AI

This app uses artificial intelligence (AI) to transcribe voice recordings with uncanny accuracy. You have to try it to believe it. Transcribe conversations or automatically turn Zoom meetings into notes.

TIMEKETTLE WT2 EDGE

These digital interpreters look like wireless earbuds, but they have a clever built-in translator with artificial intelligence (AI). They can translate dozens of languages instantly, to the other earbud or using your phone screen and speaker.

SIGNIA ACTIVE PRO

This is a new style of hearing aid for the AirPods generation. They are great for those with mild to moderate hearing loss, enhancing speech in noise. They pair with digital devices much like Bluetooth earbuds, too.

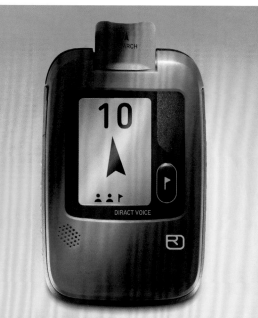

ORTOVOX DIRACT VOICE

If everyone in your off-piste ski party carries an avalanche transceiver, you will find each other in an emergency. The Diract Voice gives searchers clear, spoken directions, saving valuable, and potentially life-saving, seconds.

AMAZON ECHO SHOW 8

A voice assistant with a screen. Use it for video calls and smart home control: "Alexa, show me who's at the door." This model has a 13-megapixel wide-angle camera that digitally pans and zooms during video calls.

HARMAN KARDON CITATION 200

A stylish, rechargeable smart speaker with a big sound. Use Wi-Fi indoors to listen in HD quality or Bluetooth outdoors to stream from a phone. Google Assistant is built in for voice control and to command your other devices.

Global gadgets

Where in the world? Consumer electronics and apps from around the globe

RASPBERRY PI, WALES
The tiny single-board computer was developed by a UK charity. Most Pis are made in Pencoed, Wales, in a factory that also produces Sony camera equipment for broadcasting and is a repair centre for Sony consumer and professional products.

BLACKBERRY, CANADA
Originally called Research in Motion, the company launched a BlackBerry pager in 1999 and then a series of popular phones adept at handling email on the move. The rise of the smartphone saw BlackBerry sales decline, but the company still makes Android phones and security software.

DUOLINGO, GUATEMALA
Guatemalan academic and entrepreneur Luis von Ahn founded reCAPTCHA, the system that lets websites tell bots from humans, then sold it to Google in 2009. He went on to co-found Duolingo, the app that democratised language learning worldwide. It grew in popularity during lockdown.

MARA, RWANDA
Mara founder Ashish J Thakkar dropped out of school at 15 and started out importing computers and peripherals. Mara smartphones are proudly made in Africa — in Rwanda and South Africa — while the Mara Foundation mentors and supports young African entrepreneurs.

LOMO, RUSSIA
The original Lomo 35mm film camera originated in Leningrad (now St Petersburg). Austrian art students discovered its beauty in the 1990s, just as digital photography was taking off, and their enthusiasm kept film alive. Lomo still makes optics but not the classic camera, while film lives on in the Lomography brand.

HYUNDAI, SOUTH KOREA
Hyundai is South Korea's second biggest-earning company after Samsung. Its flagship Ioniq 5 electric car is giving Tesla and co a run for their money, thanks to a stunning design and fast charging. Hyundai has also bought robot maker Boston Dynamics. One to watch.

STOREDOT, ISRAEL
The Israeli innovator StoreDot is revolutionising electric vehicles (EVs). It has developed extreme fast-charging (XFC) lithium-ion batteries, which can be charged in five minutes, by replacing the graphite in the anode with semiconductor nanoparticles. An end to EV 'range anxiety' is in sight.

BREVILLE, AUSTRALIA
The kitchen electricals brand was founded in Sydney in 1932. It's perhaps best known in the UK for its sandwich toasters but the brand also owns Sage, the premium kitchen marque for foodies made famous by its partnership with Michelin-starred chef Heston Blumenthal.

More than a phone

A good smartphone is the centre of your tech world. Just add some gadgets to make the most of it

1 MOPHIE POWERSTATION WIRELESS XL
A discreet portable power bank, fabric-covered and the size of a phone. Charge via Lightning port or super-fast USB-C PD. Lay your phone or earbuds on it to charge wirelessly with Qi at the same time.

2 SHIFTCAM VIDEOGRAPHY KIT
Clip these lenses onto any smartphone for spectacular filmmaking. The 1.33x anamorphic lens gives footage a cinematic widescreen feel with lens flare, while the 60mm telephoto brings your subject closer for a portrait.

3 REALME GT
This flagship killer Android phone is mid-priced yet boasts a top Snapdragon 888 5G processor, a 120Hz Super AMOLED screen for extra sharpness, cinematic Dolby Atmos sound and great triple cameras. Plus, it charges from 0 to 100 per cent in 35 minutes. Big brands, be scared.

4 MCLEAR RINGPAY
The RFID chip in this ceramic smart ring lets you pay contactless in shops and tap in to public transport. You set it up via an app and top up the balance. It's waterproof, never needs charging and feels like magic.

5 BOWERS & WILKINS PI7
Bluetooth with Qualcomm aptX Adaptive means these high-end earbuds can stream high-resolution sound wirelessly. Three mics per ear handle calls, voice assistants and noise cancellation. They look as good as they sound.

6 EXPED ZIPSEAL 4
Even so-called waterproof phones can't deal with swimming, saltwater or sand. So protect your device from the beach, or driving rain, with this case. It's double-sealed against the elements but you can still use the phone.

7 GARMIN LILY
Smart watches don't have to be big beasts. This elegant version, designed by women for women, displays smartphone notifications and tracks your fitness. Choose the widgets (mini apps) it displays to tailor it to your needs.

8 JBL CHARGE 5
JBL is known for its professional PA systems and this 40W Bluetooth speaker doesn't let the brand down. It's a great party speaker, indoors or out: waterproof, with a remarkable punchy sound that isn't distorted or dominated by bass.

9 ZHIYUN SMOOTH-Q3
This three-axis gimbal turns your phone into a filmmaking camera. Taking smooth video footage is easy. It features pro filmmaking features like dolly zoom and auto follow, and includes a fill light. Useful for filmmakers, vloggers and TikTok users alike.

LE GEEK,

TECH BUZZWORDS EXPLAINED SO YOU UNDERSTAND THEM, YOU CAN EXPLAIN THEM AND YOU CAN DECIDE WHETHER TO INVEST IN THEM. REMEMBER US WHEN YOU MAKE YOUR FIRST BILLION, WILL YOU?

SO CHIC

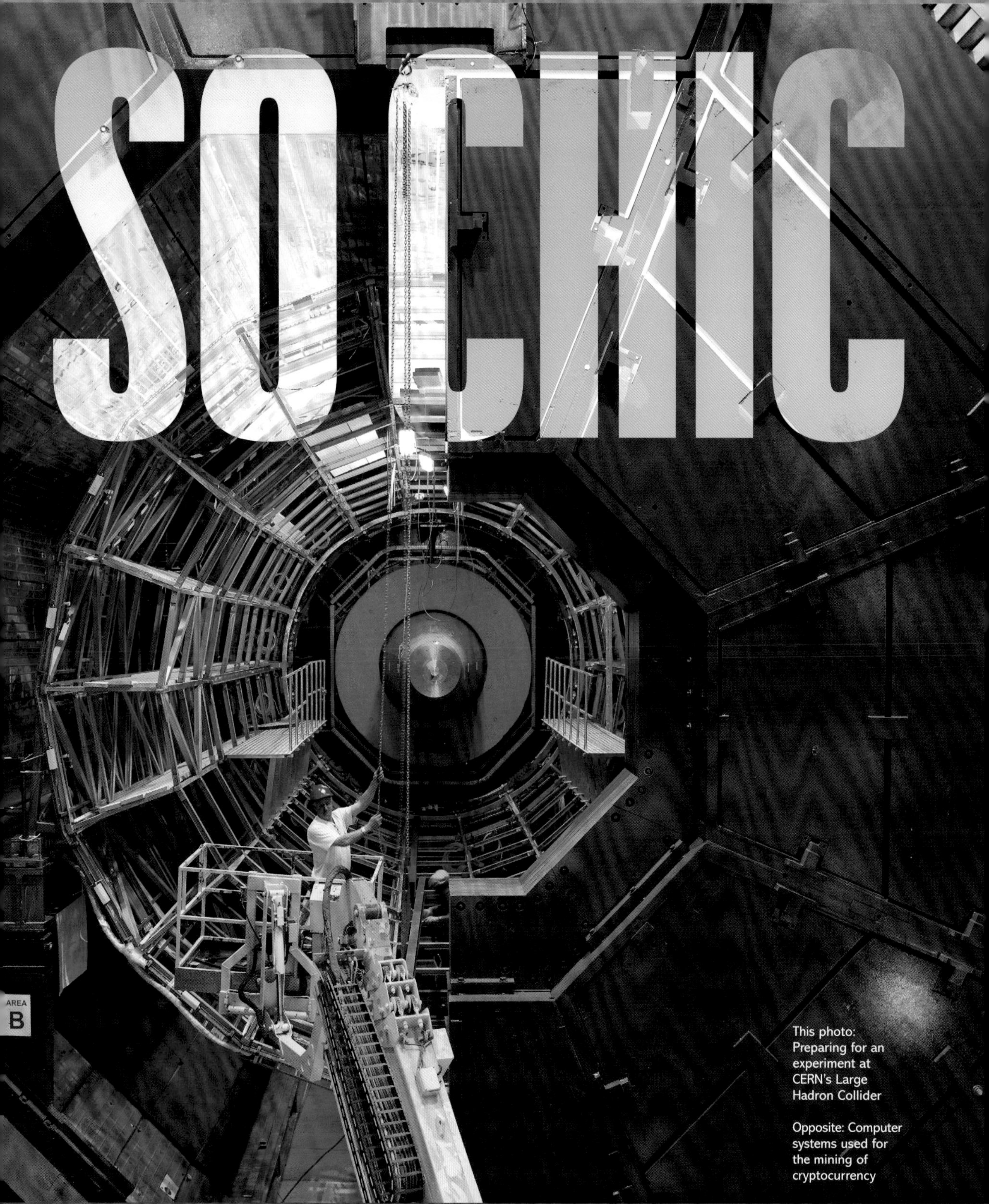

AREA
B

This photo:
Preparing for an
experiment at
CERN's Large
Hadron Collider

Opposite: Computer
systems used for
the mining of
cryptocurrency

Racks of 'mining rigs' used to maintain Bitcoin's blockchain

LARGE HADRON COLLIDER

The Large Hadron Collider (LHC) is the world's largest and most powerful particle accelerator. It's a 27km ring of superconducting magnets and accelerators, located 100 metres underground beneath the France-Switzerland border near Geneva, and is operated by CERN (the European Organisation for Nuclear Research).

CERN is the biggest particle physics laboratory in the world. It's a marvel of both science and international cooperation. It's also where the British scientist Sir Tim Berners-Lee invented the world wide web in 1989.

Inside the accelerator, two protons or ions travel around the circle in opposite directions in separate beam pipes (tubes kept at ultra-high vacuum) at close to the speed of light. Thousands of superconducting magnets maintained at very low temperatures direct them, then another magnet nudges them slightly so that they collide. It's an incredibly difficult process.

The LHC has been instrumental in discovering subatomic particles, such as the Higgs boson. The work is costly – hence the international cooperation – but it could transform our understanding of the universe, for example discovering dark matter.

BLOCKCHAIN AND CRYPTOCURRENCY

We think of money as the folding paper in our wallets, but most of the world's money exists only as numbers on screens. It doesn't exist physically. The numbers in a banking app, for example, represent the idea of money. Cryptocurrencies are a borderless alternative which are not issued by a government. Bitcoin was the first and is still the biggest cryptocurrency.

When you buy or sell Bitcoin via a digital wallet app, the transaction is recorded in a blockchain, a database that acts as a giant ledger of transactions. Complex maths is used to verify new blocks of data and add them to the chain. Lots of copies of the blockchain are kept publicly all over the world, which makes blockchain very secure: if someone hacked one copy, it would be noticed elsewhere.

Cryptocurrency mining farms use powerful computers to solve maths problems to verify each block of the blockchain. Succeed and you're paid in cryptocurrency. Mining is most profitable in places where electricity is cheap. The energy use is why it's criticised on environmental grounds.

Cryptocurrencies have seen huge growth since Bitcoin was started in 2009, but they've also been far more volatile than traditional currencies.

The Large Hadron Collider, the world's largest scientific instrument, sits 100 metres underground

EVERYDAYS: THE FIRST 5000 DAYS BY BEEPLE SOLD AT CHRISTIE'S FOR $69 MILLION, SETTING A WORLD RECORD FOR DIGITAL ART

NON-FUNGIBLE TOKENS

A fungible item can be exchanged with an equivalent. A five-pound note is fungible. A non-fungible token (NFT) cannot; it's unique. This buzzword is used to describe the owning of 'original' digital artwork. Think of it as owning the *Mona Lisa* when everyone else can only own prints. Yours is the 'original'; others have copies. Having the NFT may not even mean you own the copyright. It's more like owning a certificate of authenticity.

Like cryptocurrencies, ownership is recorded on a blockchain that can't be forged. But also, like cryptocurrencies, NFTs are an investment and their value could go up or down. The only surefire way to make money seems to be actually making the art and selling it as an NFT.

NFT sales have included a 10-year-old animated image, or GIF, of a cat (fetching nearly $600,000) and artwork by the musician Grimes (totalling around $6 million). The NFT *Everydays: The First 5000 Days* by the graphic designer Beeple sold at Christie's for a staggering $69,346,250, setting a record for digital art. It is a huge collage of one picture a day taken over 13 years. Expect the NFT trend to trickle down to affordable collectables, like Panini stickers or Pokémon cards.

GRAPHENE

Graphene, a flat lattice of carbon atoms, was first truly isolated in 2004 by researchers pulling one-atom-thin layers from a block of graphite using ordinary sticky tape. They won the Nobel Prize in Physics for their discovery. The supermaterial is so thin that it's transparent. It's also insanely strong, ridiculously light, conducts heat and electricity, and is 100 times stronger than steel of the same weight.

Gadgets featuring graphene include performance fabrics, speaker and headphone drivers, batteries and supercapacitors. Graphene can revolutionise drug delivery and medical sensors and can even be used to make aeroplane wings lighter. The material is expensive but is getting cheaper.

The reversible Vollebak Graphene Jacket, which has one side coated in graphene, can boost body temperature by around 2°C when the graphene side is used as a lining. Or wear it graphene side out to soak up energy from the sun, then reverse it to harness that energy. Vollebak describes the experimental jacket as "the very first step towards our end goal of creating bionic clothing that is both bulletproof and intelligent".

Above: Raw graphite from which ultra-light graphene is extracted

Right: Vollebak's Graphene Jacket is partly made from a graphene supermaterial with remarkable physical properties

JON EXPLAINS FUEL CELLS
The Gadget Show's JON BENTLEY talks about a green alternative to petrol

Fuel cells are, in some ways, similar to batteries; both have an anode, a cathode and an electrolyte. But instead of them being charged up, fuel cells need a constant supply of fuel for a reaction to happen, together with a catalyst. They are a reality. In fact, they've been around for over 150 years. They're used in things like forklift trucks and they're good in space.

Natural gas is used to create the hydrogen that vehicles use now, but it could be environmentally cleaner. There's a filling station in Rotherham, South Yorkshire, that uses electricity generated by a wind turbine for electrolysis to split water into its constituent parts: hydrogen and oxygen. Hydrogen made in this way is potentially a carbon-neutral fuel.

A quantum computer can solve problems faster than any other machine

QUANTUM COMPUTING

Traditional computers are binary systems: each bit of data is a one or a zero. But a quantum bit (qubit) can store more data because it can be a one, a zero or an infinite number of values in between. It can even represent multiple values at once. 'Quantum entanglement' means that qubits can work together, so computing power grows exponentially when you add more.

Physics works differently at a subatomic scale, so the business bit of a quantum computer is tiny. Liquified helium cools it down to a fraction above absolute zero. At that temperature, tiny superconducting circuits in the chip have quantum properties, and we can work with qubits.

Traditional computers function methodically with logic but they are slow to solve big problems. Quantum computers deal with uncertainty and probabilities, so they are adept at considering all the possible solutions to a problem simultaneously.

We've already reached quantum supremacy where quantum computers can solve problems that would take traditional computers thousands, if not millions, of years. But they're delicate and so, so pricey. Don't expect one in your smartphone any time soon.

MRNA

Messenger RNA (mRNA) is a single strand of ribonucleic acid. It's a sequence of genetic code that tells cells what proteins to build. Its single-strand shape is not to be confused with DNA's double helix.

mRNA is in the news because it's the technology used in both the Pfizer-BioNTech and Moderna Covid-19 vaccines. The vaccines' mRNA give your cells instructions to make the S (spike) protein pieces found on the surface of the Covid-19 virus. (If you visualise the classic grey-and-red image of the Covid-19 virus, the S proteins are the red-coloured spiky bits on the outside.)

Your body creates antibodies to the S protein. The mRNA (which never enters the nucleus of your cells) degrades but the antibodies stay. So, if you are later infected with the Covid-19 virus, the antibodies are ready to fight the virus. This gives you a huge head start in fighting the disease.

In the future, mRNA-based therapies could be used in the same way to trigger an immune response to other diseases including cancer.

Strands of mRNA are used in the manufacture of Pfizer and Moderna Covid-19 vaccines

Joined-up thinking

These gadgets are the glue that binds our technology. They connect with all the other gadgets and converge to control our smart tech

LENOVO THINKPAD X1 FOLD

The world's first PC with a folding screen. Use it folded like a mini laptop with a keyboard, unfold it into a tablet with a 13.3-inch OLED touchscreen, or open it like a book.

GET CONNECTED ⊖

REMARKABLE 2

Do away with paper notebooks. This 4.7mm-thin E Ink tablet combines the digital and handwritten, so you can turn penned notes into digital text, sketch ideas and write comments directly on PDFs.

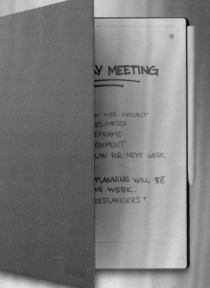

EMOTIV EPOC X

This lightweight wireless headset offers 14-channel EEG for monitoring brain activity, plus 10-axis motion sensors to detect head movement. It's aimed at research but can be used to mind-control tech.

LYNX SPECIAL EDITION

This mixed reality (MR) headset has twin cameras on the front for 3D video passthrough. Toggle between VR (virtual reality) and AR (augmented reality) to choose your level of immersion in the virtual world.

NETGEAR NIGHTHAWK XR1000

A flagship high-speed router designed for gaming and juggling the demands of intensive use and family life.

OMNICHARGE OMNI 20+

A mega charger for all tech, not just the USB-powered stuff. It's surprisingly small but has a large capacity and both DC and AC outputs. Plug in your devices just like plugging into a wall socket.

TICKRMETER

This small ticker sits on a desktop and tracks investments in real time. Its E Ink display can update frequently without being power-hungry. You pick the stock (or cryptocurrency or forex) to follow and set alerts.

PANASONIC ENELOOP BQ-CC65

We fret about the environment but use disposable batteries. This charger can refresh old rechargeables and counts how many disposables you've saved. Use it with nickel-metal hydride cells, such as Eneloop batteries that come pre-charged and are slow to discharge.

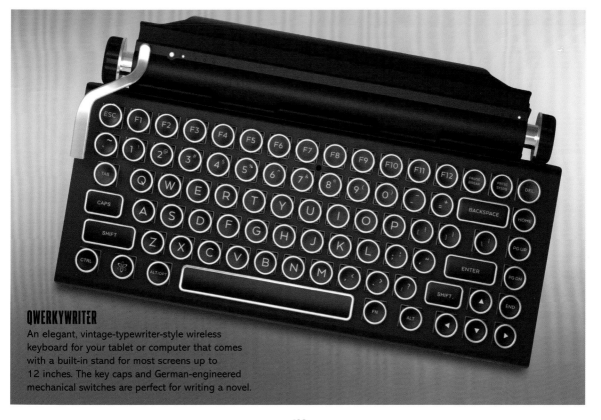

QWERKYWRITER

An elegant, vintage-typewriter-style wireless keyboard for your tablet or computer that comes with a built-in stand for most screens up to 12 inches. The key caps and German-engineered mechanical switches are perfect for writing a novel.

Credits

EDITORIAL
Edited by Caramel Quin
Creative direction and design
 by Anton Jacques
Subedited by John Lewis and
 Fiona Russell
Picture research by Tom Broadbent

Contributors:
Georgie Barrat
Jon Bentley
Craig Charles
Ortis Deley
Chris Haslam
Paul Henderson
John Lewis
Caramel Quin
Harry Wallop
Rob Waugh
Jordan Erica Webber

With thanks to North One and Channel 5

"5" is a trademark of Channel 5
Broadcasting Ltd

ILLUSTRATIONS
Cover artwork: Lisa Sheehan

Presenter photography: cover, pp2/3,
pp30–1, p37, pp39–40, p68, p71, p75,
pp77–8, p84, pp108–9, p115, pp117–9,
p124, p132, p147, p153, p155–7, p163,
p169, p179, p184 David Titlow

Product photography: pp24–5, pp62–3,
pp102–3, pp134–5, pp178–9 Pixeleyes
Photography

Illustration: pp64–6, p68, pp80–3,
p85 Fernando Volken Togni

Dude, where's my flying car?: pp10/11
Benoit Lebel; p12 Courtesy of Tesla,
Inc.; p13 Alamy; p14 Lamborghini; p15
Waymo; p16 Historic Vehicle Association;
p17 Joby Aviation © Bradley Wentzel;
JetPack Aviation

Starman waiting in the sky: pp22/23
SpaceX

Fast vs slow: p26 Pierre Mangez; p27
Zoonar GmbH; p28 Virgin Hyperloop,
agefotostock; p29 Alamy, Hybrid Air
Vehicles; p31 Volkswagen

Craig Charles… the joker: p38 Alamy;
p39 BBC Photo Library, Alamy;
p40 Alamy; p41 The Gadget Show –
North One

Magic moments: pp42–7 Neil Edwards;
pp44–7 The Gadget Show – North One

Would you like to play a game?: pp48–9
Activision Publishing, Inc. (Activision,
Call of Duty, Call of Duty Black Ops,
Call of Duty Warzone and Warzone are
trademarks of Activision Publishing, Inc.
All other trademarks and trade names
are the properties of their respective
owners), Pokémon Go, Among Us, Roblox,
Dreams, Playdate – Saturday Edition,
Rocksmith+, A Short Hike, Butterfly
Soup, Ratchet & Clank – Rift Apart,
Soup Pot, Sea of Thieves; pp50/51
The Gadget Show – North One; p51
Michal Konkol/Riot Games, Playdate;
p53 Colin Young-Wolff/Riot Games; p54
The Gadget Show – North One; p55
Joe Brady/Excel Esports

Cathedral of books: pp60/61 Getty
Images

Mini marvels: p67 Raspberry Pi,
Ben Q, Harry Wallop; p69 Musik –
och teatermuseet, The National Trust
Photolibrary, Alamy, Sorin Mares,
MiNe, Alamy, Miguelon756-5303,
Astell & Kern

Georgie Barrat… the thinker: pp76–7,
p79 The Gadget Show – North One;
p79 Sleep Life – Georgie Barrat

Make-do and mend: p83 Dualit; p85
Harry Wallop

Performance enhancers: p86 Pulseroll;
pp86–7 AJ Boxing/Getty Images; p88
Maja Hitij/Getty Images, Whoop Strap;
p89 TAG Heuer; p90 Alex Livesey/Getty
Images, Bromley Sports; p91 Cameron
Smith/Getty Images, Pulseroll; p92 Dave
Benett/Getty Images, VanMoof; p93 Syo
van Vliet/Red Bull Content Pool, Thermic;
p94 Gareth Cattermole/UEFA/Getty
Images, Nike; p95 Wout Beel/Deceuninck
– Quick-Step Team; Oakley

Your future could be in cyber:
pp100–1 Alamy; p100 Omkaar Kotedia;
p101 Dexcom, Dani Clode Design,
Osseointegration.eu, Alamy

Sleep, the final frontier: 104–5 Adobe
stock; p106 Metronaps; p107 Getty
Images, Kokoon; p109 Bose, Google Nest

Ortis Deley… the action man: p116, p119
The Gadget Show – North One; p117 ITV/
Rex Shutterstock

Science fiction, science fact!: pp120/121
Moviestore Collection Ltd/Alamy;
p122 Mike Seyfang, Healthy.io, CBS
Photo Archive/Getty Images, Sony;
p123 Lexus Slide; AF archive/Alamy;
p124 BBC Photo Library, Time Kettle,
Everett Collection Inc/Alamy; p125 TCD/
Prod.DB/Alamy, Oblong Technologies,
Alamy, Amazon

The New Normal: pp126–7 Moley
Robotics; p128 Simplisafe; p129 Maria
Cruseman, Switchbot, Harry Wallop,
Vaha; p130–1 Bridger Automation; p131
Angi.com, Alamy; p133 Velux, Dyson,
Lightair Ionflow

Seed capital: pp136–7 Getty Images

Bathroom 2.0: pp142-3 Hansgrohe
Raintunes; p144 Kaldewei; p145
SensoWash; Simplehuman; p146
Crosswater, Grohe; p147 Alamy

Jon Bentley… the boffin: p154, p157
The Gadget Show – North One; p154
BBC Photo Library; p157 Alamy

Rise of the robots: pp158–9 Tomohiro
Ohsumi/Getty Images, NASA; p160 Getty
Images, DARPA; p161 The Gadget Show
– North One, NASA; p162 Honda Miimo
HRM 70 Live, Airwheel, Samsung; p163
Getty Images

One thing to rule them all: pp164/165
Adobe Stock; p167 Steve Jurvetson;
p168 Getty Images, Harry Wallop; p169
Microsoft; p170 Samsung Sero; p171
Alamy, Ruben de Rijcke

Global gadgets: p176 Getty Images,
Duolingo

Le geek, so chic: p180 Getty Images;
p181 © 2006 CERN – Maximilien Brice;
p182 Getty Images, © 2017 CERN –
Maximilien Brice; p183 © Christie's
Images Limited 2021; p184 Vollebak/
Sun Lee; p185 Adobe Stock, Alamy

Published by Black Dog Press Limited, a company registered
in England and Wales with company number 11182259.
Black Dog Press Limited is an imprint within the SJH Group.
Copyright is owned by the SJH Group. All rights reserved.

Black Dog Press Limited
The Maple Building
39–51 Highgate Road
London NW5 1RT
United Kingdom

+44 (0)20 8371 4047
office@blackdogonline.com
www.blackdogonline.com

Printed in the UK by Kingsbury Press

ISBN 978-1-912165-35-3

British Library in Cataloguing Data. A CIP record for this book is
available from the British Library.

The SJH Group would like to thank our colleague
Garry Blackman to whom we dedicate this book